Microbial Fuel Cells 2018

Microbial Fuel Cells 2018

Special Issue Editor
Jung Rae Kim

MDPI • Basel • Beijing • Wuhan • Barcelona • Belgrade

Special Issue Editor
Jung Rae Kim
Pusan National University
Republic of Korea

Editorial Office
MDPI
St. Alban-Anlage 66
4052 Basel, Switzerland

This is a reprint of articles from the Special Issue published online in the open access journal *Energies* (ISSN 1996-1073) in 2018 (available at: https://www.mdpi.com/journal/energies/special_issues/microbial_fuel_cells_2018)

For citation purposes, cite each article independently as indicated on the article page online and as indicated below:

LastName, A.A.; LastName, B.B.; LastName, C.C. Article Title. *Journal Name* **Year**, *Article Number*, Page Range.

ISBN 978-3-03921-535-5 (Pbk)
ISBN 978-3-03921-534-8 (PDF)

© 2019 by the authors. Articles in this book are Open Access and distributed under the Creative Commons Attribution (CC BY) license, which allows users to download, copy and build upon published articles, as long as the author and publisher are properly credited, which ensures maximum dissemination and a wider impact of our publications.

The book as a whole is distributed by MDPI under the terms and conditions of the Creative Commons license CC BY-NC-ND.

Contents

About the Special Issue Editor vii

Preface to "Microbial Fuel Cells 2018" ix

Jiseon You, John Greenman and Ioannis Ieropoulos
Novel Analytical Microbial Fuel Cell Design for Rapid in Situ Optimisation of Dilution Rate
and Substrate Supply Rate, by Flow, Volume Control and Anode Placement
Reprinted from: *energies* 2018, *11*, 2377, doi:10.3390/en11092377 1

Dong-Mei Piao, Young-Chae Song and Dong-Hoon Kim
Bioelectrochemical Enhancement of Biogenic Methane Conversion of Coal
Reprinted from: *energies* 2018, *11*, 2577, doi:10.3390/en11102577 13

Jiyun Baek, Changman Kim, Young Eun Song, Hyeon Sung Im, Mutyala Sakuntala and
Jung Rae Kim
Separation of Acetate Produced from C1 Gas Fermentation Using an Electrodialysis-Based
Bioelectrochemical System
Reprinted from: *energies* 2018, *11*, 2770, doi:10.3390/en11102770 26

Paweł P. Włodarczyk and Barbara Włodarczyk
Microbial Fuel Cell with Ni–Co Cathode Powered with Yeast Wastewater
Reprinted from: *energies* 2018, *11*, 3194, doi:10.3390/en11113194 38

Ki Nam Kim, Sung Hyun Lee, Hwapyong Kim, Young Ho Park and Su-Il In
Improved Microbial Electrolysis Cell Hydrogen Production by Hybridization with a TiO_2
Nanotube Array Photoanode
Reprinted from: *energies* 2018, *11*, 3184, doi:10.3390/en11113184 47

Priyadharshini Mani, Vallam Thodi Fidal Kumar, Taj Keshavarz, T. Sainathan Chandra and
Godfrey Kyazze
The Role of Natural Laccase Redox Mediators in Simultaneous Dye Decolorization and Power
Production in Microbial Fuel Cells
Reprinted from: *energies* 2018, *11*, 3455, doi:10.3390/en11123455 60

About the Special Issue Editor

Jung Rae Kim is an Associate Professor in the School of Chemical and Biomolecular Engineering at Pusan National University, Korea. He received his BS and MS degrees in Chemical Engineering at Pusan National University, Korea, and his Ph.D. in Environmental Engineering at Pennsylvania State University, USA in 2006 with a thesis on microbial fuel cells. Then, he moved into the Sustainable Environment Research Centre (SERC), Faculty of Advanced Technology in University of Glamorgan (University of South Wales at present), United Kingdom. He has conducted a UK national EPSRC Supergen Biological fuel cell project as a research fellow since 2006, and as a permanent post senior research fellow since 2010. He joined the School of Chemical and Biomolecular Engineering in Pusan National University (PNU) in September 2012 and opened the Bioenergy and Bioprocess Engineering Lab at PNU. His main research aim is the development of sustainable bioelectrochemical systems for bioenergy and useful chemical production. Recently he has focused on novel bioconversion bioprocesses using bioelectrochemical systems: 1,3-PDO and 3-HP production from glycerol, and electrosynthesis, an electrode-based C1 gas (CO_2/CO/CH_4) conversion into intermediary metabolites. He also has expertise in microbial fuel cell system fabrication and operation for applications in bioenergy and biorefinery processes. He has published over 90 SCI(E) research papers with more than 5000 citations (h-index: 31).

Preface to "Microbial Fuel Cells 2018"

Microbial fuel cells (MFCs) convert an organic substrate into electricity using micro-organisms as a biocatalyst. The concept of MFCs was introduced in the early twentieth century. However MFC has been extensively examined since 1990's. Various applications with MFC concepts has been suggested and implemented in biosensor and wastewater treatment etc. The publication and citation for MFC has been increased to reflect the recent interest in sustainable and renewable bioenergy. The control of interface between live cell bacteria and carbon electrode as well as MFC system development, are identified as key factors for further improvement of system performance. In this respect, This Special Issue of Energies explore the latest developments in MFC technology.

Specifically, this book encompass:
- In situ optimization of important parameters of MFC
- Application of MFC concept for methane conversion of coal
- MFC type electrodialysis for volatile fatty acids separation
- Alternative cathode electrode of MFC for wastewater treatment application
- A MFC for hydrogen production by hybridization with a TiO2 Nanotube
- A MFC for dye decolorization with a natural laccase redox mediator

Jung Rae Kim
Special Issue Editor

Article

Novel Analytical Microbial Fuel Cell Design for Rapid in Situ Optimisation of Dilution Rate and Substrate Supply Rate, by Flow, Volume Control and Anode Placement

Jiseon You *, John Greenman and Ioannis Ieropoulos *

Bristol BioEnergy Centre, University of the West of England, Bristol BS16 1QY, UK; john.greenman@uwe.ac.uk
* Correspondence: jiseon.you@uwe.ac.uk (J.Y.); ioannis.ieropoulos@brl.ac.uk (I.I.); Tel.: +44-117-328-6318 (I.I.)

Received: 6 August 2018; Accepted: 30 August 2018; Published: 9 September 2018

Abstract: A new analytical design of continuously-fed microbial fuel cell was built in triplicate in order to investigate relations and effects of various operating parameters such as flow rate and substrate supply rate, in terms of power output and chemical oxygen demand (COD) removal efficiency. This novel design enables the microbial fuel cell (MFC) systems to be easily adjusted in situ by changing anode distance to the membrane or anodic volume without the necessity of building many trial-and-error prototypes for each condition. A maximum power output of 20.7 ± 1.9 µW was obtained with an optimal reactor configuration; 2 mM acetate concentration in the feedstock coupled with a flow rate of 77 mL h^{-1}, an anodic volume of 10 mL and an anode electrode surface area of 70 cm^2 (2.9 cm^2 projected area), using a 1 cm anode distance from the membrane. COD removal almost showed the reverse pattern with power generation, which suggests trade-off correlation between these two parameters, in this particular example. This novel design may be most conveniently employed for generating empirical data for testing and creating new MFC designs with appropriate practical and theoretical modelling.

Keywords: microbial fuel cell (MFC); anode distance; anodic volume; flow rate; dilution rate; substrate supply rate; treatment efficiency; power generation

---

1. Introduction

In the past two decades, scientific interest in the microbial fuel cell (MFC) technology has increased rapidly. Direct conversion of organic matter including various types of waste into electricity is one key aspect that enable this technology to stand out among other renewable energy related technologies. Moreover, its application is not limited to only electrical energy generation and waste treatment. For instance, the same working principles applied to the MFC technology can be used, with the supply of external power, for producing useful products such as hydrogen [1,2], acetate [3,4], methane [5,6] as well as desalinate water [7,8]. Resource recovery and bio-sensing [9–13] are also highly active fields in the MFC research. Along with practical development of the technology, microorganisms involved in electricity generation have drawn a great deal of attention too. Electrochemically active biofilms (EABs) on MFC electrodes continue to be studied to better understand the anodic biofilm properties involved in substrate digestion, utilisation and transformation of chemicals, all resulting electricity generation. EABs have also been used for microbial computing [14–16].

It is generally agreed that continuous flow bioreactors for either planktonic or biofilm culture systems are more efficient than their corresponding batch culture processes in terms of start-up, turnaround, maintenance, efficiency and control. Small-scale MFC using perfusable anode electrodes are particularly suited for continuous operation since biofilms form on a highly porous material, ensuring that diffusion limitation (of substrate to biofilm) does not limit growth and that following monolayer saturation of the electrode (termed "mother layer"), all new daughter cells that form are shed and washed away by laminar flow of bulk medium at high liquid shear rates. The electrode-attached cell population (the biofilm) remains as a constant number of cells with time and with constant flow, the biofilm quickly reaches dynamic steady state [17]. As a biofilm system, the small-scale perfusion anode MFC is analogous to a chemostat system in terms of steady state, and therefore the effects of flow rate, chamber volume and feedstock concentration can be more easily determined, and well-known terms used in chemostat theory (e.g., Monods equations) applied in modelling.

In this study, a novel analytical MFC design was developed, which enables the system to be easily set, tuned or adjusted to a given condition by altering reactor configurations such as anode position or reactor volume. With the help of this novel MFC design, this study aimed to: (1) demonstrate the effects of anode chamber volume and distance to anode electrode as important parameters in reactor configuration in terms of electricity generation, and (2) investigate the relationships between flow rate, volume, dilution rate and substrate supply rate on power output and COD reduction. In addition to these findings, which validate the new MFC design, potential applications of this analytical MFC can be used for (1) analytical studies, (2) MFC modelling, and (3) enabling new MFC designs with specific target purposes.

2. Materials and Methods

2.1. Microbial Fuel Cell Design

For this study, a disposable polypropylene 50 mL syringe (Terumo, UK) was used as an MFC chassis, in order to change the anodic volume readily using a plunger without disturbing the anodic microbial community. The barrel of syringe was used as the anodic chamber after cutting off the tip; this left a 32 mm diameter open window. A cation exchange membrane (CMI-7000, Membrane International Inc., Ringwood, NJ, USA) was placed at this end and a hot-pressed activated carbon cathode, prepared as previously described [18] with a total surface area of 8.0 cm^2 (diameter: 32 mm) was placed onto the membrane; this cathode was open to air. A laser cut acrylic ring (thickness: 3 mm) was mounted on the tip in order to hold both the membrane and cathode. Plain carbon fibre veil (PRF Composite Materials, UK) 70 cm^2 total area, with a folded projected area of 2.9 cm^2) was used as an anode electrode. A 15 cm nickel-chrome wire threaded through the anode came out the back of the syringe, which facilitated moving the anode inside the anodic chamber. This design allowed a maximum anodic chamber volume of 50 mL (taking into account the displacement volume of the electrode) and all tests were carried out in triplicates. A detailed schematic of the design of this syringe MFC is shown in Figure 1. All the outlets were sealed with the exception of a single outlet appropriate for the chosen volume.

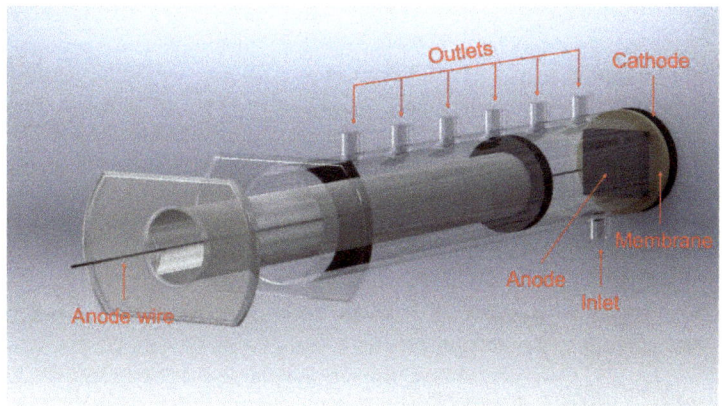

Figure 1. Computer Aided Design (CAD) image of a syringe microbial fuel cell (MFC) used in this study.

2.2. Inoculum, Feedstock and Operation

Sewage sludge from a local wastewater treatment plant (Wessex Water, Saltford, UK) was used to inoculate the MFCs, after being enriched with 1% tryptone and 0.5% yeast extract. During the first week, 10 mL of synthetic wastewater [19] was provided as the feedstock on a daily basis. Subsequently, the batch type of feedstock supply was switched to continuous feeding mode, using a 16-cahnnel peristaltic pump (205U, Watson Marlow, Falmouth, UK) with variable flow rates, ranging from 19.2 mL h^{-1} to 306.9 mL h^{-1}. The synthetic wastewater was prepared by adding the following to 1 L of distilled water: 0.270 g (NH$_4$)$_2$SO$_4$, 0.060 g MgSO$_4$·7H$_2$O, 0.006 g MnSO$_4$·H$_2$O, 0.50 g NaCl, 0.13 g NaHCO$_3$, 0.003 g FeCl$_3$·6H$_2$O, 0.006 g CaCl$_2$·2H$_2$O, 0.006 g K$_2$SO$_4$. Sodium acetate was used as the carbon energy source at variable concentrations, ranging between 0.1 mM and 4.0 mM.

Throughout the work, a 1.5 kΩ external load was connected to each MFC, which was determined based on polarisation runs (data not shown) that were carried out at the start of the experiments. Power output of the MFCs was monitored in real time in volts (V) against time using an ADC-24 Channel Data Logger (Pico Technology ltd., St Neots, UK). All experiments were carried out in a temperature-controlled environment, at 22 ± 2 °C, and repeated at least 3 times.

2.3. Anode Placement Test

In order to investigate power and COD removal related to the distance between anode and membrane (or cathode since it was directly attached to the obverse side of the membrane), the MFC reactor was set to its maximum volume of 50 mL thus the anode was able to be moved to give adjustment between 0 up to 6 cm from the membrane. For this test, 2 mM of acetate was supplied at a flow rate of 19 mL h^{-1}, which resulted in dilution rate of 0.38 h^{-1} and nutrient supply rate of 0.04 mmol h^{-1}. The dilution rate (D) is inversely related to the hydraulic retention time (HRT); where $HRT = 1/D$. The dilution rate was calculated by dividing the flow rate (f) (how much medium flows into the vessel per hour) by the chamber volume (V), since $D = f/V$.

The substrate supply rate (R) is defined by:

$$R = S \times f$$

where S is the substrate molar concentration (mmol L^{-1}) and f is feedstock flow rate (L h^{-1}).

2.4. Substrate Supply Rate and Dilution Rate Test

For this set of experiments, three variables (feedstock flow rate, feedstock concentration and MFC reactor volume) were set to determine a range of substrate supply rates and dilution rates. When a variable changed, the other two variables were fixed. Tested ranges of flow rate and concentration were 19–307 mL h^{-1} for feedstock flow rate and 0.1–4.0 mM for feedstock acetate concentration. For these tests, the anodic volume was set to 30 mL. The effects of changing the anodic volume were studied by changing the volume of the anodic chamber from 10 mL to 50 mL in 10 mL increments. Feedstock concentration and flow rate were fixed at 2.0 mM and 38 mL h^{-1} respectively. During all these tests, the anode was located next to the membrane, thus the distance between the anode and membrane was designated as 0 cm. Each concentration of feedstock or anodic volume condition was set for at least 2 h, which was long enough for MFCs to reach a stable level of power output.

2.5. Chemical Oxygen Demand (COD) Analysis

Influent and effluent were collected from the feedstock storage tank and individual MFCs respectively and samples analysed for COD. The potassium dichromate oxidation method (COD LR test vials; Camlab Ltd., Cambridge, UK) [20] and a photometer (Lovibond MD 200; The Tintometer Ltd., UK) were used to determine COD values of each sample. Efficiency of COD removal was calculated as E_{COD} (%) = $(COD_{IN} - COD_{OUT})/COD_{IN} \times 100$, where COD_{IN} (mg L^{-1}) is the influent COD and COD_{OUT} (mg L^{-1}) is the effluent COD.

3. Results

3.1. Effect of Anode Distance from the Membrane

As shown in Figure 2, power output decreased, and COD removal rate increased as the anode moved a greater distance from the membrane. Power decrease with increasing distance between the two electrodes is most likely due to the longer traveling distance for protons to the membrane, thus higher ohmic losses [21,22]. The optimum distance between the anode and the membrane for power output was 1 cm, where the power output was 4.8% higher than when the anode was in contact with the membrane; arguably this is within the error margin of readings between 0 cm and 1 cm, but possibly the result of oxygen crossover through the membrane to the anode (for the 0 cm condition). However, at distances between 2–6 cm there is a decreasing trend of power output, clearly showing that these distances are sub-optimal.

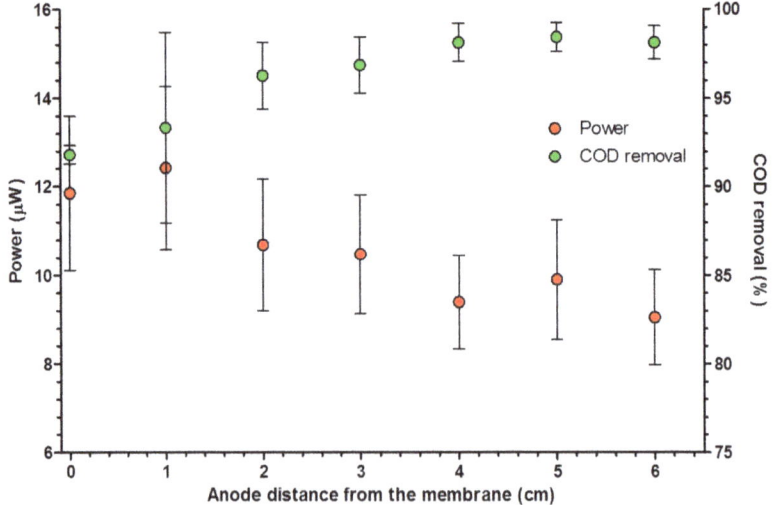

Figure 2. Power output and chemical oxygen demand (COD) removal efficiency with different distance between anode and membrane.

For treatment efficiency of each configuration tested, the distance of 5 cm showed the highest removal rate of 98.4 ± 1.4%, although this is not significantly different to the values recorded for 4 cm, or 6 cm. The high removal rate may again be explained by the influence of oxygen diffusion from the cathode, allowing cells in the planktonic phase as well as those in the biofilm to continue utilising the organic substrates, lower the COD, but also competing with, or inhibiting the metabolism of anodophiles, thus reducing electrical output. The overall COD reduction was over 90% in all cases, probably due to the relatively low nutrient supply rate (0.04 mmol h^{-1}) and moderate dilution rate (0.38 h^{-1}) employed during this experiment.

3.2. Effect of Dilution Rate

Different dilution rates ranging from 0 to 10.2 h^{-1} were tested by changing feedstock flow rate (19–307 mL h^{-1}), concentration (0.1–4 mM), and anodic volume (10–50 mL). Figure 3 describes relations between dilution rate and power output, and COD reduction rate. Previous work has confirmed that following a moderate period of time in batch culture, once beyond the decline phase, the power output of all MFCs eventually drops to zero, in line with the theoretical principle that a supply rate of zero fuel will eventually give zero metabolism. At low substrate concentrations (0.5 mM, green line in Figure 3), the power output remained low, but measurable across all dilution rates, including the highest tested D (10.2 h^{-1}) with a steady state value around 0.13 µW. At a higher concentration of carbon energy (C/E) source (1.0 mM), a relationship can be seen between increasing dilution rate and increasing power until it reaches a limit at a dilution rate of 5.1 h^{-1}, where the power plateaus at about 4 µW for any higher dilution rates. Similar patterns of behaviour (power increases with increasing D until a plateau is reached) are also observed at higher concentrations of substrate. The power output then remains the same despite further increases in the dilution rate.

At low concentrations of C/E (0.5 and 1.0 mM), growth is strongly limited by lack of fuel (C/E limiting condition), even when supplied at a high flow or dilution rate. Also, it is likely that a significant proportion of the C/E fuel is required for maintaining microbial cell functions (maintenance energy). At higher concentrations of C/E, for example 2.0 and 4.0 mM, the maximum power output reaches levels between 15 and 16 µW, and the maintenance energy becomes a much smaller proportion of the total energy output. It should be noted that doubling the concentration from 2.0 to 4.0 mM had no observable effect in producing additional electrical power showing that C/E concentrations are growth limiting at or just below 2.0 mM. At lower fuel concentrations (e.g., between 0 and 1 mM) the power output is strongly dependent on dilution rate, suggesting that the C/E is most probably limiting growth and metabolic rate and thus power generation.

Figure 3 (vertical black line with points) also shows the results of measuring the energy outputs obtained for a range of nine different concentrations of acetate (from 0.1 to 4.0 mM), but at a constant flow rate/dilution rate (D = 10.2 h^{-1}). At this high and constant D, the effect of C/E alone on power output was again clearly observed. For low concentrations of C/E ranging 0.1 mM and 1.0 mM, power output increased from 0.0 µW to 5.6 µW, whereas there was no significant increase in power output when higher concentrations of C/E (between 2.0 mM and 4.0 mM) were used.

On the other hand, COD reduction, which reflects substrate utilisation, decreased as the dilution rate increased for all tested conditions (Figure 3B). At the lowest dilution rate of 0.6 h^{-1}, COD reduction rate was between 81.9 and 100%. Then, it went down to 11.1–50.6% at the highest dilution rate of 10.2 h^{-1}. In all continuous biofilm flow systems there is a portion of C/E that will flow around the electrode and not be utilised by the microbial cells and this will be higher for higher substrate concentrations and/or higher flow and dilution rates.

Although the maximum tested dilution rate of 10.2 h^{-1} was thought to be quite high, the detrimental effect of liquid shear rate causing cell detachment was not observed, which suggests that the biofilms on the electrode are very strongly attached and resilient to shear force removal.

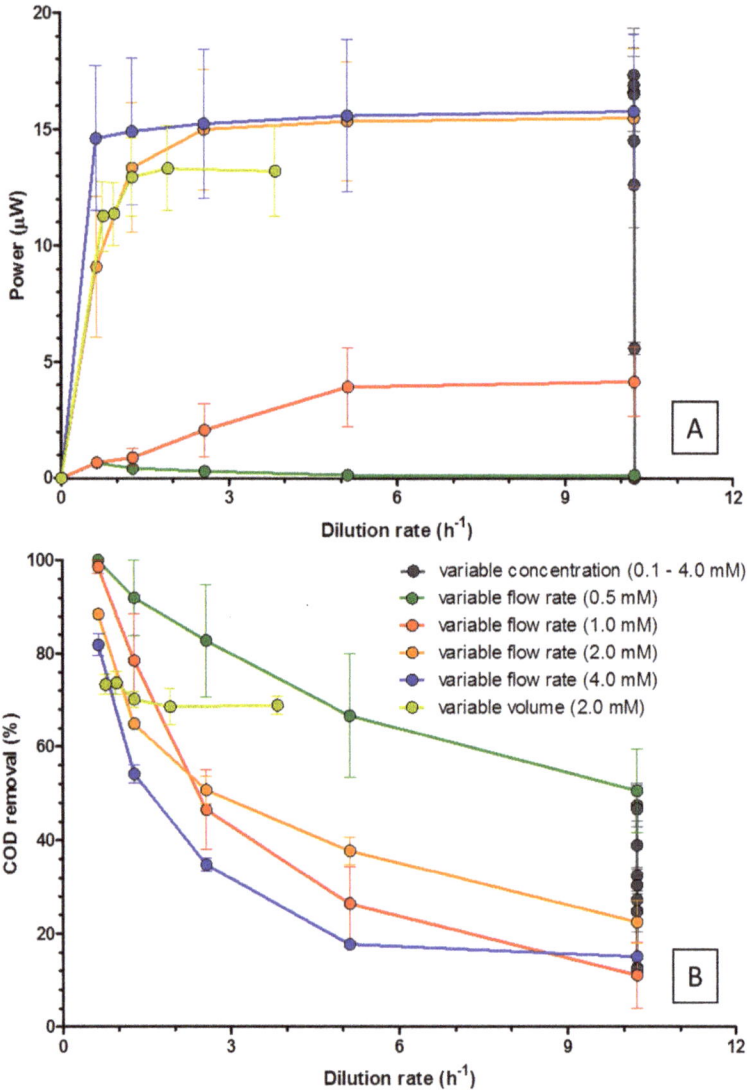

Figure 3. Power output (**A**) and COD removal efficiency (**B**), at different dilution rates.

3.3. Effect of Substrate Supply Rate

Figure 4 shows power production and COD removal rate, subjected to variation in nutrient supply rate (*R*), using the same data above. For low substrate concentration of 0.5 mM (green line), the maximum nutrient supply rate was only 0.15 mmol h^{-1} even at the highest dilution rate of 10.2 h^{-1}. At a higher concentration of C/E (1.0 mM), power increased when *R* increased up to 0.12 mmol h^{-1}, then there was no further increase beyond this point, suggesting that the power output is directly proportional to the saturation fraction of the uptake system, which is given by $S/(K_m + S)$, where K_m is the Michaelis constant.

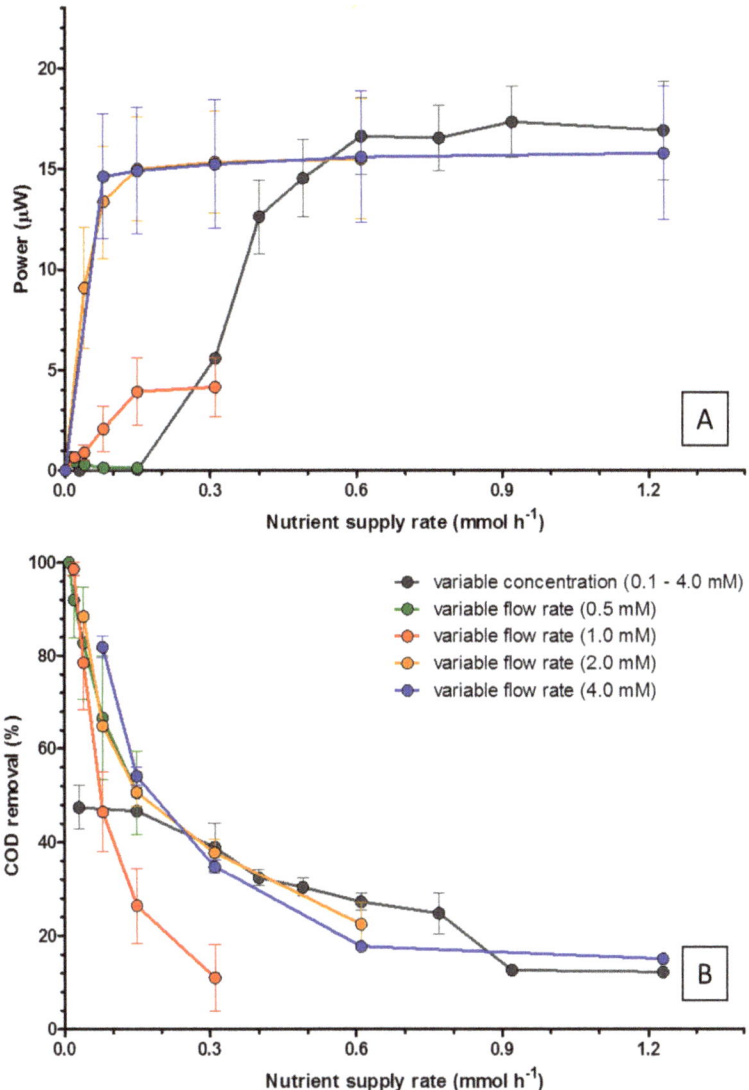

Figure 4. Power output (**A**) and COD removal efficiency (**B**), at different nutrient supply rates.

In general, the COD removal rate decreased with increasing nutrient supply rates. At very low nutrient supply rates (0.01–0.02 mmol h^{-1}), COD reduction rates were over 90%, which suggests that most of the C/E source was fully utilised for cell growth and maintenance. The COD reduction rate, then decreased at higher supply rates as previously described for effects of dilution rate.

3.4. Perfusion Anode Biofilm and Quasi Steady State

Unlike the planktonic mode of bacterial growth and existence (e.g., as in the case of a chemostat), biofilms are associated with two types of populations, attached cells that are firmly bound and remain at constant populations and the planktonic phase (detached cells washing out). For biofilms, a steady state occurs when growth accumulation is matched by loss of cells from the system and such biofilms can be maintained in quasi-steady state for as long as the operational factors of the system such as feedstock composition, nutrient supply rate and dilution rate are kept constant. Figure 5 shows stable power outputs (steady states) produced by triplicate MFCs over seven days demonstrating that the replicate MFC units are highly reproducible.

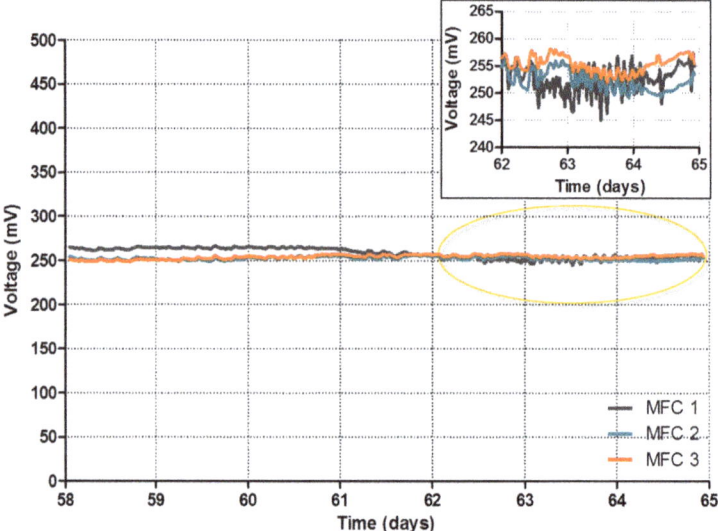

Figure 5. Stable power generation over seven days, from triplicate MFCs under metabolic steady state. Inset graph presents a magnified view of the highlighted period.

The Monod model is most commonly used to describe the growth kinetics of cells growing in steady state. Figure 6 describes the response of the MFCs in terms of power output and COD utilisation towards changes in substrate concentration. From these data, it is possible to calculate the half-rate saturation constant (Ks value) of 1.114 mM.

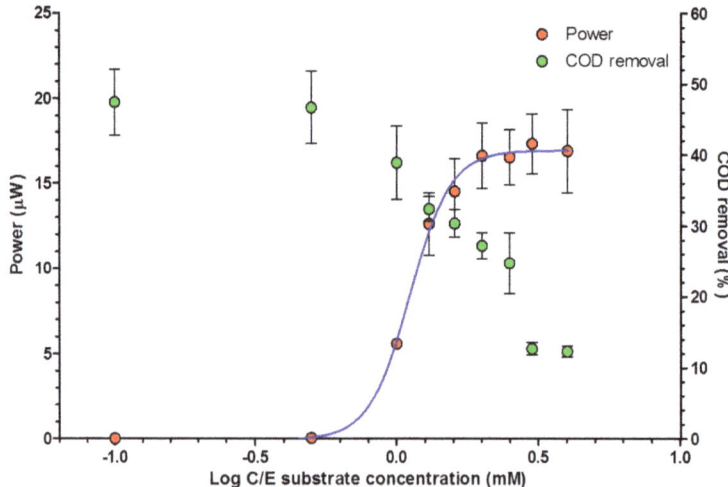

Figure 6. Power generation and COD reduction response to substrate concentration at high dilution rate ($D = 10.2$ h^{-1}).

3.5. Effect of Anodic Volume on Power and COD Reduction

Since anodic volume is usually considered as a fixed design element, controlling dilution rate or hydraulic retention time (*HRT*) is done through changing the flow rate. However, the novel design of MFCs used in this study enables the anodic volume to change, it can be also an operating parameter.

Power output decreased, whereas COD removal efficiency increased with increasing the anodic volume (corresponding D decreased from 3.8 h^{-1} to 0.8 h^{-1}). Power density normalised by the anodic volume shows an even clearer trend opposite to the anodic volume increase as shown in Figure 7. This indicates that under the given parameters such as fixed size of both electrodes and membrane, flow rate, feedstock concentration and electrode spacing, the smallest anodic volume of 10 mL was the best value for maximum power generation. For maximum COD removal, the biggest volume of 50 mL achieved the best output. These results are in accordance to those reported by others [23,24]; a shorter *HRT* contributes to a decrease of COD removal. In this test, however, *HRT* was controlled by changing the anodic volume instead of flow rates. Although change in planktonic bacterial population is negligible in this test, due to the relatively short time of each volume condition, it can also have an effect on MFC power output and COD removal. Larger anodic chamber volumes provide greater space for planktonic bacteria to grow, thus higher total bacterial population. This does not necessarily contribute to power generation, but consumes more organic matter in the feedstock thus achieving a higher COD removal. If the substrate is complex in terms of its molecular structure, a larger volume would be preferable since fermentative heterotrophs can break the substrate down first, making it more easily available for the anodic biofilm community.

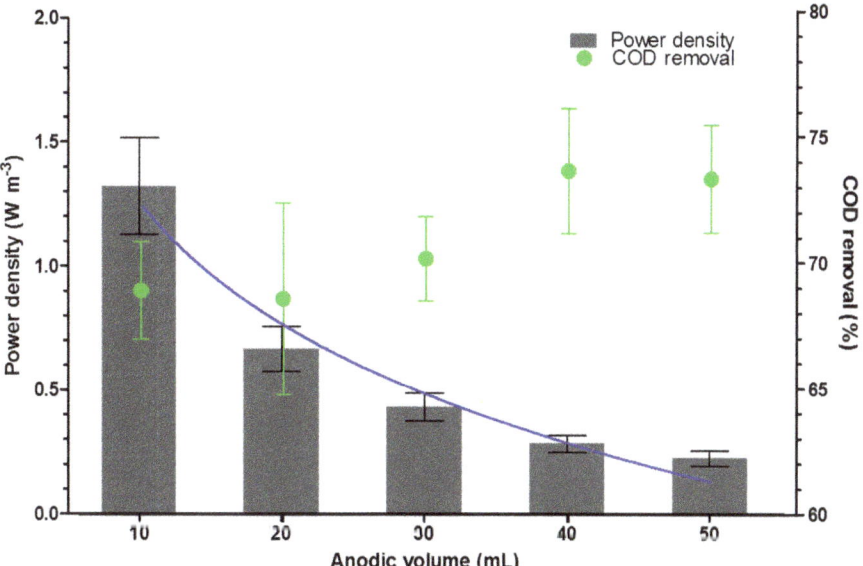

Figure 7. Volumetric power density and COD removal efficiency with different anodic volume (blue line shows the trend of volumetric power density).

4. Discussion

MFCs with thick diffusion-limiting biofilms have a low growth rate/metabolic rate (thus slow response time) because of slow diffusion of substrate from the medium through the thick biofilm to the inner conductive layers. Such biofilms typically form over non-porous (solid) electrodes, especially in batch culture where mechanism of electron transfer is via mediators as much as it is by direct conduction methods. This can be minimised in small-scale MFCs, by using highly perfusable electrodes and high flow rate. In these conditions, the biofilm remains thin. The MFC is not limited by diffusion and soluble mediator is rapidly washed away, so it is only the thin biofilm in direct conductance that produces rapid responses and can grow at maximum specific growth rate.

The continuous flow model allows biofilms to grow and reach dynamic steady states, where the attached cell population continues to grow, metabolise and thus produce electrical power, and yet the perfusable biofilm remains constant (i.e., non-accumulating) over time by shedding of new daughter cells [17]. The relatively high flow rates and high dilution rates employed (e.g., $D = 10.2$ h^{-1}) did not seem to affect the stability of the biofilm (although this has not yet been determined by use of molecular approaches at the ecological level). However, it can be concluded that for the highest C/E-excess conditions (substrate concentration of 4 mM), a higher flow rate produced no significant change in electrical power output, which remained at maximum power (15–16 µW), even at very high flow rates (equivalent to $D = 10.2$ h^{-1}). For lower concentration of substrate (2.0 mM) increasing the flow rate (from 19.2 to 76.7 mL h^{-1}) gave increasing power output up to a maximum (15 µW), which then remained the same despite further increases in flow rate. The same pattern was observed when lower substrate concentration of 1.0 mM was tested. Increases in the flow rate increased the power output up to a dilution rate of $D = 5.1$ h^{-1}, where the maximum power for this concentration of substrate (3.9 µW) was obtained. Therefore, it can be concluded that if the C/E supply rate is growth limiting, then the power can be maximised by increasing the flow rate. It also suggests that Fick's laws of diffusion do not need to be incorporated into a mathematical model of the biological behaviour of such biofilm-electrodes. These findings are useful when considering the advantages of cascades and optimising the flow rate down such cascades.

Another important finding was the effect of anode working volume on power output and COD reduction efficiency. As can be seen in Figure 7, when normalised for anodic working volume, higher power was generated from the smaller volume (10 mL) than the larger (50 mL) and with an inverse relationship for the tested volumes in between; this is in line with previous reports [25,26]. Treatment (COD) efficiency showed the opposite, and although the percentage reduction varied between 69–74% for all tested parameters, higher COD reduction was recorded for the larger volume experiments; this may have been the result of the fixed flow rate and fixed substrate concentration chosen for this line of experiments, and should therefore be further investigated under different fixed conditions.

5. Conclusions

Novel MFC design allows in situ placement of anode and its distance apart from the membrane-cathode to be optimised. The design is particularly suited for observing the effects of changes in the physicochemical conditions, particularly concentration of C/E in the feedstock, flow rate and thus the supply rate and dilution rate of the system, on metabolism of the anodic biofilm and thus power output. Moreover, this novel design would help to create new design of MFCs by comparing the performance in terms of power generation and treatment efficiency under different operating conditions. It would also be useful for MFC modelling to help better understand the technology. Building a truly tenable MFC system can be achievable with auxetic material as an anode and chassis. Future work needs to seek suitable materials for electrode and chassis.

Author Contributions: Conceptualization, J.G. and I.I.; Formal analysis, J.Y.; Funding acquisition, I.I.; Investigation, J.Y.; Methodology, J.Y. and J.G.; Project administration, J.Y.; Visualization, J.Y.; Writing—original draft, J.Y.; Writing—review & editing, J.G. and I.I.

Funding: This research was funded by the Engineering and Physical Sciences Research Council (EPSRC) UK, grant number EP/N005740/1. The APC was funded by the Research Councils UK (RCUK) Open Access Block Grant, available through the University of the West of England, Bristol.

Conflicts of Interest: The authors declare no conflict of interest.

References

1. Kadier, A.; Simayi, Y.; Abdeshahian, P.; Azman, N.F.; Chandrasekhar, K.; Kalil, M.S. A comprehensive review of microbial electrolysis cells (MEC) reactor designs and configurations for sustainable hydrogen gas production. *Alexandria Eng. J.* **2016**, *55*, 427–443. [CrossRef]
2. Escapa, A.; Mateos, R.; Martínez, E.J.; Blanes, J. Microbial electrolysis cells: An emerging technology for wastewater treatment and energy recovery. From laboratory to pilot plant and beyond. *Renew. Sustain. Energy Rev.* **2016**, *55*, 942–956. [CrossRef]
3. Jiang, Y.; Su, M.; Zhang, Y.; Zhan, G.; Tao, Y.; Li, D. Bioelectrochemical systems for simultaneously production of methane and acetate from carbon dioxide at relatively high rate. *Int. J. Hydrogen Energy* **2013**, *38*, 3497–3502. [CrossRef]
4. Xafenias, N.; Mapelli, V. Performance and bacterial enrichment of bioelectrochemical systems during methane and acetate production. *Int. J. Hydrogen Energy* **2014**, *39*, 21864–21875. [CrossRef]
5. Villano, M.; Monaco, G.; Aulenta, F.; Majone, M. Electrochemically assisted methane production in a biofilm reactor. *J. Power Sources* **2011**, *196*, 9467–9472. [CrossRef]
6. Babanova, S.; Carpenter, K.; Phadke, S.; Suzuki, S.; Ishii, S.; Phan, T.; Grossi-Soyster, E.; Flynn, M.; Hogan, J.; Bretschger, O. The Effect of Membrane Type on the Performance of Microbial Electrosynthesis Cells for Methane Production. *J. Electrochem. Soc.* **2017**, *164*, H3015–H3023. [CrossRef]
7. Carmalin Sophia, A.; Bhalambaal, V.M.; Lima, E.C.; Thirunavoukkarasu, M. Microbial desalination cell technology: Contribution to sustainable waste water treatment process, current status and future applications. *J. Environ. Chem. Eng.* **2016**, *4*, 3468–3478. [CrossRef]
8. Al-Mamun, A.; Ahmad, W.; Baawain, M.S.; Khadem, M.; Dhar, B.R. A review of microbial desalination cell technology: Configurations, optimization and applications. *J. Clean. Prod.* **2018**, *183*, 458–480. [CrossRef]
9. Gajda, I.; Greenman, J.; Melhuish, C.; Santoro, C.; Li, B.; Cristiani, P.; Ieropoulos, I. Electro-osmotic-based catholyte production by Microbial Fuel Cells for carbon capture. *Water Res.* **2015**, *86*, 108–115. [CrossRef] [PubMed]
10. You, J.; Greenman, J.; Melhuish, C.; Ieropoulos, I. Electricity generation and struvite recovery from human urine using microbial fuel cells. *J. Chem. Technol. Biotechnol.* **2016**, *91*. [CrossRef]
11. Xiao, Y.; Zheng, Y.; Wu, S.; Yang, Z.-H.; Zhao, F. Nitrogen recovery from wastewater using microbial fuel cells. *Front. Environ. Sci. Eng.* **2016**, *10*, 185–191. [CrossRef]
12. Brunelli, D.; Tosato, P.; Rossi, M. Microbial fuel cell as a biosensor and a power source for flora health monitoring. In Proceedings of the 2016 IEEE Sensors, Orlando, FL, USA, 30 October–3 November 2016; pp. 1–3.
13. Chouler, J.; Cruz-Izquierdo, Á.; Rengaraj, S.; Scott, J.L.; Di Lorenzo, M. A screen-printed paper microbial fuel cell biosensor for detection of toxic compounds in water. *Biosens. Bioelectron.* **2018**, *102*, 49–56. [CrossRef] [PubMed]
14. Greenman, J.; Ieropoulos, I.; McKenzie, C.; Melhuish, C. Microbial computing using Geobacter electrodes: output stability and consistency. *Int. J. Unconv. Comput.* **2006**, *2*, 249–265.
15. Greenman, J.; Ieropoulos, I.; Melhuish, C. Biological computing using perfusion anodophile biofilm electrodes (PABE). *Int. J. Unconv. Comput.* **2008**, *4*, 23–32.
16. Tsompanas, M.-A.I.; Adamatzky, A.; Sirakoulis, G.C.; Greenman, J.; Ieropoulos, I. Towards implementation of cellular automata in Microbial Fuel Cells. *PLOS ONE* **2017**, *12*, e0177528. [CrossRef] [PubMed]
17. Ledezma, P.; Greenman, J.; Ieropoulos, I. Maximising electricity production by controlling the biofilm specific growth rate in microbial fuel cells. *Bioresour. Technol.* **2012**, *118*, 615–618. [CrossRef] [PubMed]
18. You, J.; Walter, X.A.; Greenman, J.; Melhuish, C.; Ieropoulos, I. Stability and reliability of anodic biofilms under different feedstock conditions: Towards microbial fuel cell sensors. *Sens. Bio-Sensing Res.* **2015**, *6*, 43–50. [CrossRef]
19. Winfield, J.; Ieropoulos, I.; Greenman, J. Investigating a cascade of seven hydraulically connected microbial fuel cells. *Bioresour. Technol.* **2012**, *110*, 245–250. [CrossRef] [PubMed]
20. Eaton, A.D.; Clesceri, L.S.; Greenberg, A.E.; Franson, M.A.H. *Standard Methods for the Examination of Water and Wastewater*; American Public Health Association: Washington, DC, USA, 1995.
21. Sajana, T.K.; Ghangrekar, M.M.; Mitra, A. Effect of pH and distance between electrodes on the performance of a sediment microbial fuel cell. *Water Sci. Technol.* **2013**, *68*, 537. [CrossRef] [PubMed]

22. Ahn, Y.; Hatzell, M.C.; Zhang, F.; Logan, B.E. Different electrode configurations to optimize performance of multi- electrode microbial fuel cells for generating power or treating domestic wastewater. *J. Power Sources* **2014**. [CrossRef]
23. Juang, D.F.; Yang, P.C.; Kuo, T.H. Effects of flow rate and chemical oxygen demand removal characteristics on power generation performance of microbial fuel cells. *Int. J. Environ. Sci. Technol.* **2012**, *9*, 267–280. [CrossRef]
24. You, S.J.; Zhao, Q.L.; Jiang, J.Q.; Zhang, J.N. Treatment of domestic wastewater with simultaneous electricity generation in microbial fuel cell under continuous operation. *Chem. Biochem. Eng. Q.* **2006**, *20*, 407–412.
25. Papaharalabos, G.; Greenman, J.; Melhuish, C.; Ieropoulos, I. A novel small scale Microbial Fuel Cell design for increased electricity generation and waste water treatment. *Int. J. Hydrogen Energy* **2015**, *40*, 4263–4268. [CrossRef]
26. Greenman, J.; Ieropoulos, I.A. Allometric scaling of microbial fuel cells and stacks: The lifeform case for scale-up. *J. Power Sources* **2017**, *356*, 365–370. [CrossRef]

© 2018 by the authors. Licensee MDPI, Basel, Switzerland. This article is an open access article distributed under the terms and conditions of the Creative Commons Attribution (CC BY) license (http://creativecommons.org/licenses/by/4.0/).

Article

Bioelectrochemical Enhancement of Biogenic Methane Conversion of Coal

Dong-Mei Piao [1], Young-Chae Song [1,*] and Dong-Hoon Kim [2]

1   Department of Environmental Engineering, Korea Maritime and Ocean University, 727 Taejong-ro, Yeongdo-Gu, Busan 49112, Korea; jingying46@kmou.ac.kr
2   Department of Civil Engineering, Inha University, 100 Inha-ro, Nam-gu, Incheon 22212, Korea; dhkim77@inha.ac.kr
*   Correspondence: soyc@kmou.ac.kr; Tel.: +82-51-410-4417

Received: 30 August 2018; Accepted: 26 September 2018; Published: 27 September 2018

Abstract: This study demonstrated the enhancement of biogenic coal conversion to methane in a bioelectrochemical anaerobic reactor with polarized electrodes. The electrode with 1.0 V polarization increased the methane yield of coal to 52.5 mL/g lignite, which is the highest value reported to the best of our knowledge. The electrode with 2.0 V polarization shortened the adaptation time for methane production from coal, although the methane yield was slightly less than that of the 1.0 V electrode. After the methane production from coal in the bioelectrochemical reactor, the hydrolysis product, soluble organic residue, was still above 3600 mg chemical oxygen demand (COD)/L. The hydrolysis product has a substrate inhibition effect and inhibited further conversion of coal to methane. The dilution of the hydrolysis product mitigates the substrate inhibition to methane production, and a 5.7-fold dilution inhibited the methane conversion rate by 50%. An additional methane yield of 55.3 mL/g lignite was obtained when the hydrolysis product was diluted 10-fold in the anaerobic toxicity test. The biogenic conversion of coal to methane was significantly improved by the polarization of the electrode in the bioelectrochemical anaerobic reactor, and the dilution of the hydrolysis product further improved the methane yield.

Keywords: coal; lignite; methane; biogenic conversion; bioelectrochemical reactor; inhibition

1. Introduction

Coal-bed methane (CBM) is an important source of natural gas that is formed in subsurface coal seams. The CBM is commonly extracted by wells, but the extraction rate is limited by the formation rate of CBM in the coal seam [1–3]. There are two types of CBMs in the coal seam, thermogenic and biogenic, that are converted from the organic matter contained in coal [3]. While the thermogenic CBM is formed as a side product of coalification at an elevated temperature and pressure, the biogenic formation of CBM is a continuous process that is carried out by a series of anaerobic microbial conversions of the organic matter in the coal. However, the rate of the microbial conversion of coal into methane is very low, and the methane yields are also too low to be economical [4,5]. The commercial availability of CBM requires improvements to the methane conversion rate and the yield from coal.

In the anaerobic conversion process of organic matter to methane, organic polymers are hydrolyzed and fermented by acidogens to intermediates, such as acetate and hydrogen. The intermediates are finally converted to methane by methanogens [6,7]. The physicochemical characteristics of the organic matter are key factors affecting the anaerobic conversion process. The organic matter contained in coal is mainly composed of hydrophobic substances, such as lignin, that are undergoing coalification, which are hydrolyzed very slowly. The hydrolysis products are composed of recalcitrant cyclic compounds, including long chain fatty acids, alkanes (C_{19}-C_{36}), and various aromatic hydrocarbons, and are difficult to for the acidogens to ferment [4,5,8]. The cyclic compounds can generally be degraded under aerobic conditions by oxidizing their rings or adding oxygen to their nuclei to open the rings [8,9]. When molecular oxygen is not available, some substances, including nitrate, iron, and sulfate, can be also used as the electron acceptor for ring opening of the cyclic compounds [8]. However, when there are no electron acceptors at a low redox potential, carbon dioxide can be reduced to methane, but the ring opening reaction is less thermodynamically favorable.

To date, the methane conversion of coal has been mainly improved by increasing the bioavailability of coal, biostimulation, and bioaugmentation. The bioavailability of coal can be increased to some extent by reducing the coal particle size, increasing the porosity, and adding surfactants [10,11]. Bioaugmentation and biostimulation, in which a microbial consortium or inorganic nutrients, such as nitrogen, phosphorus, trace elements, and vitamins, are supplied to the coal bed, have also been used effectively for promoting coal conversion to methane [2,3,12,13]. However, the methane yield obtainable from 1 g of coal is still only a few tens of μL to a few mL [2,5,11,14]. The organic content of coal varies depending on the coalification degree, and in the case of lignite, is 0.5–0.8 g COD (chemical oxygen demand) per g of coal. In anaerobic degradation of organic matter, theoretically, 1 g of COD can be converted to 350 mL of methane. This suggests that the conversion potential of coal to methane is fairly high.

The free energy change driving the redox reaction at the electrode surface depends on the polarized potential of the electrode. Recently, the principle of an electrochemical redox reaction on an electrode surface has been applied to improve the anaerobic degradation of organic matter [7,15]. An anaerobic reactor with a polarized electrode is called a bioelectrochemical anaerobic reactor. In a bioelectrochemical anaerobic reactor, the electroactive microorganisms, including exoelectrogenic fermentation bacteria (EFB) and electrotrophic methanogenic archaea (EMA), are enriched on the surfaces of the polarized electrode [15–17]. The electroactive microorganisms can donate or accept electrons to the outside of the cell through the cytochrome C or conductive pili that extend to the outer membrane of the cell [7,18]. EMA are microorganisms that produce methane by directly accepting the electrons from the EFB and then reduce carbon dioxide [15,16]. Recently, it has been revealed that EFB and EMA can be enriched not only on the electrode surface, but also in the bulk solution [15,18]. The electrons can be directly transferred between the interspecies of the EFB and EMA through the electrode, conductive materials, and direct contact [18–20]. In the anaerobic degradation of organic matter, the limitations of the kinetics and thermodynamics are considerably mitigated by methane production via direct interspecies electron transfer (DIET) [18,21]. This suggests that the bioelectrochemical approach has great potential to improve the methane conversion of coal. However, the bioelectrochemical methane conversion of coal has not yet been studied.

In this study, we first demonstrated that the polarized electrode remarkably improves the methane conversion of coal in a bioelectrochemical anaerobic reactor. The rate-limiting step controlling the overall methane conversion of coal was estimated and the inhibitory effects of the hydrolysis products of coal on the methane production were also evaluated.

2. Materials and Methods

2.1. Coal and Seed Sludge

Commercially available Canadian lignite as coal was purchased from a local distributor (Aquajiny Co., Daegu, Korea). The percentage of volatile solids in the lignite was 28.5%, and the organic and moisture contents were 0.52 g COD/g lignite and 18.4%, respectively. For the anaerobic batch experiment converting coal to methane, the lignite was powdered by crushing with a mortar and pestle, and screened with a 1 mm sieve, followed by drying at 105 °C for 12 h. The medium for the anaerobic batch experiment was prepared with the initial concentrations of 2.45 g/L $NaH_2PO_4 \cdot 2H_2O$, 4.58 g/L $Na_2HPO_4 \cdot 12H_2O$, 0.31 g/L NH_4Cl, and 0.31 g/L KCl. Small amounts of vitamins and trace metals were also added to the medium, following a previously reported method [22]. Anaerobic sludge collected from a sewage treatment plant (Busan, South Korea) was screened with a 1 mm sieve and then used as an inoculum by precipitating in a refrigerator at 4 °C for 24 h. The initial pH of the inoculum for the anaerobic batch experiment and the anaerobic toxicity test were 7.17 and 7.25, respectively, the VSS (volatile suspended solids) were 13.4 and 16.1 g/L, and the alkalinities were 2114 and 3702 mg/L as $CaCO_3$.

2.2. Experimental Apparatus for the Anaerobic Conversion of Coal to Methane

The bioelectrochemical anaerobic batch reactor (2 sets, effective volume 0.5 L, diameter 8.5 cm, and height 10 cm) was prepared using a cylindrical acrylic resin tube (Figure 1). The top of the anaerobic batch reactor was covered with acrylic plate and joined with a flange to seal the reactor. A blade for mixing was placed inside the reactor. The blade was connected to a DC motor over the acrylic cover plate using a steel shaft. The sampling ports for the gas and liquid and an off-gas valve were installed at the acrylic cover plate. The sampling ports for the gas and liquid were covered with n-butyl rubber stoppers. The liquid sampling port and the steel shaft hole of the acrylic cover plate were sealed with acrylic tubes extending into the liquid phase. The off-gas valve was connected to a floating-type gas collector by a rubber tube. The gas collector was filled with acidic saline water to prevent dissolution of the biogas. Copper foils (0.3 T, copper 99.9%, KDI Co., Seoul, Korea) with a large area (26 cm × 9 cm) and a small area (5.5 cm × 7 cm) were prepared. The surfaces of the foils were coated with a dielectric polymer (alkydenamel, VOC 470 g/L, Noroo paint Co., Busan, Korea) and used as a pair of electrodes. The electrodes were installed at the inner wall of the reactor and the outer wall of the sealing tube for the steel shaft. The interval between the inner and outer electrodes was 3.3 cm. The electrodes were connected to the terminals of an external voltage source (ODA Technologies, CO., Incheon, Korea) with titanium wires.

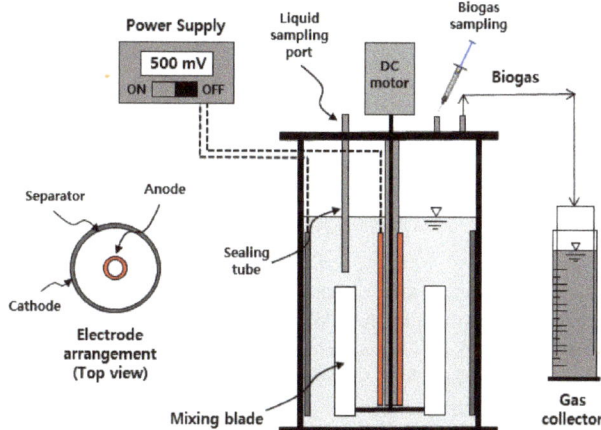

Figure 1. Schematic diagram of the bioelectrochemical anaerobic batch reactor.

For the anaerobic batch experiments, seed sludge (250 mL), medium (250 mL), and lignite (2.5 g) were added to the reactor. The electrodes were polarized by applying voltage of 1 or 2 V using the external voltage source. The anaerobic batch reactor with an applied voltage of 1 V was called as BEF1 and the reactor with 2 V was called BEF2. The anaerobic batch reactor was placed in a temperature-controlled room (35 °C) and mixed with the rotating the blade at 120 rpm. An anaerobic batch reactor without an applied voltage was operated under the same conditions as a control. An anaerobic batch reactor without added lignite was prepared to examine the methane production from the inoculum alone.

2.3. Anaerobic Toxicity Test

The hydrolysis product of the lignite was collected from the anaerobic batch reactor after the experiment. The COD and pH of the hydrolysis product were 4.16 g/L and 8.25, respectively. In the anaerobic toxicity test of the hydrolysis product, serum bottles of 125 mL were prepared. The inoculum (40 mL) and the hydrolysis product solution, with volumes ranging from 8 to 32 mL, were added to the serum bottles, and the anaerobic medium was filled to a total liquid volume of 80 mL. As an easily degradable substrate, 0.2 g of glucose was also added to the serum bottle. The serum bottle was flushed with nitrogen gas and sealed with rubber and aluminum caps using a crimper. The sealed serum bottle was incubated in a rotary shaker (120 rpm) at 35 °C. A serum bottle without the hydrolysis product in the liquid was also incubated under the same conditions and used as the control. Another serum bottle with added inoculum and medium was used as a blank to correct for the methane production from the seed sludge. All the anaerobic toxicity tests in the serum bottles were performed in triplicate.

2.4. Analysis and Calculation

During the anaerobic batch experiments, small amounts of liquid sample was intermittently collected, and the physicochemical properties, such as TS, VS, TCOD, SCOD, and VSS, were analyzed according to the Standards Method. The pH was measured with a pH meter (YSI pH1200 laboratory pH meter 115–230 V (T1)). At the end of the anaerobic batch experiment, VFAs (volatile fatty acids) in the liquid sample were analyzed by HPLC (high-performance liquid chromatography, UltiMate 3000, Thermo Scientific, Sunnyvale, USA) equipped with a UV detector and a separation column (Amines HPX-87H). The biogas production was monitored using a floating-type gas collector, and the biogas composition (methane, hydrogen, and carbon dioxide) was analyzed using a GC (gas chromatograph Clarus 580, PerkinElmer Co., Ltd. Shelton, USA) equipped with a thermal conductivity detector and a separation column (Porapak-Q, 6 ft × 1/8", SS). The cumulative production volumes of methane and hydrogen were estimated from the biogas production and the biogas composition using Equation (1).

$$V_{CH_4/H_2} = C_{CH_4/H_2} \times (V_{RH} + V_{GT} + V_{GC}) \tag{1}$$

where V_{CH_4/H_2} is the cumulative production volume of methane or hydrogen (mL) and C_{CH_4/H_2} is the percentage of methane or hydrogen in the biogas. V_{RH} is the head space of the batch reactor, V_{GT} is the volume of the rubber tube between the reactor and the gas collector, and V_{GC} is the gas phase volume in the gas collector. The methane and hydrogen production volumes were converted to the corresponding values at standard temperature and pressure (STP), following previous studies [15,22]. At the end of the anaerobic batch experiment, a cyclic voltammetry experiment for the bulk solution was performed using a potentiostat (ZIVE SP1, WonA Tech, Seoul, Korea) at the scan rate of 10 mV/s in the voltage range of −1.0 to 1.0 V. Small pieces of stainless steel mesh (1 cm × 1 cm) were used as the working and counter electrodes, and an Ag/AgCl electrode was used as the reference electrode. The redox peak potential and the current were determined from the cyclic voltammogram obtained using the SmartManager software (ZIVE BP2 Series, WonA Tech, Seoul, Korea).

In the anaerobic toxicity test, the biogas production was intermittently monitored using a lubricated glass syringe, and the mean methane production was obtained from the biogas volume and the composition, following a previous study [23], then it was converted to the corresponding value at STP. The cumulative methane production was fit with the Modified Gompertz equation, Equation (2), to estimate the lag phase time, the maximum methane production rate, and the ultimate methane production.

$$P = P_u \times \exp[-\exp(\mu_m \times \exp(1) \times (\lambda - t)/P_u + 1)] \tag{2}$$

where P is the cumulative methane production (mL) at time t, P_u is the ultimate methane production (mL), μ_m is the maximum methane production rate (mL/day), and λ is the lag phase time (day). The nlstools package in R was used for the fitting the cumulative methane production.

3. Results and Discussion

3.1. Bioelectrochemical Conversion of Coal to Methane under Electrostatic Field

In BEF1 with an applied voltage of 1 V, the methane production was delayed after start-up, but suddenly increased to 32.5 mL/g lignite on the 24th day. During the following several days, the cumulative methane production tended to decrease somewhat, but increased again to 52.5 mL/g lignite on the 31st day (Figure 2). To the best of our knowledge, this is the highest value of methane yield of coal reported to date. In previous studies, one of the highest methane yields of coal was 7.4 mL/g lignite, depending on the type of coal and degree of coalification [3,5]. This was mainly attributed to the organic matter in the coal, which was not a good substrate to be metabolized easily by microorganisms. The common organic constituents in coal are complex organic polymers, such as lignin, which are difficult to decompose [2]. There have been several attempts to improve the methane conversion of coal. The methane yield can be improved to 1.66–2.93 mL/g subbituminous coal by adding nutrients, such as yeast, algae, and cyanobacteria [3]. Aerobic pretreatment that promoted the bioavailability of coal improved the methane yield to 4.98 mL/g lignite [2]. However, the methane yield obtained in BEF1 was 10.5–31.6 times higher than those reported in previous studies [2,3]. Intriguingly, the methane production in BEF1 was discontinuous, resembling a pulse (Figure 3). In anaerobic digestion, the common intermediates are acetate and hydrogen, which are converted to methane by acetoclastic and hydrogenotrophic methanogens, respectively [6,21]. During the operation of BEF1, hydrogen was also observed in the biogas. The amount of hydrogen in the biogas increased and decreased repeatedly. However, the methane production in BEF1 did not exhibit the same behavior as that of hydrogen (Figure 3), indicating that there was little correlation between the hydrogen consumption and the methane production. This implies that the methane conversion mechanism of coal in BEF1 may be different from the known interspecies hydrogen transfer. It has been revealed that the homoacetogens oxidize carbon dioxide with hydrogen in a bioelectrochemical system to produce acetate, a substrate for exoelectrogens [24–26]. In BEF1, it seems that the acidogens fermented the hydrolysis product to produce hydrogen, and the homoacetogens produced acetate from the hydrogen and carbon dioxide. In a bioelectrochemical anaerobic reactor, the exoelectrogens on the electrode surface or in the bulk solution fermented the low molecular organic matter and released electrons to outside the cell [7,15]. The electrotrophic methanogens take the electrons directly to produce methane by reducing carbon dioxide [15,27]. In the cyclic voltammogram for the bulk solution in BEF1, two pair of redox peaks were observed at $-0.014/-0.239$ V vs. Ag/AgCl ($E_f = -0.127$ V) and 0.913/0.025 V vs. Ag/AgCl ($E_f = 0.461$ V) (Figure 4). The redox peaks in the voltammogram indicate the presence of electroactive substances, such as electroactive microorganisms or electron transfer mediators in the bulk solution [15]. In anaerobic digestion, when the redox potential is more negative than -0.44 V vs. Ag/AgCl, carbon dioxide can be thermodynamically reduced to methane at standard conditions. This suggests that the 2nd redox peaks are not likely to be involved in the methane conversion of coal. The formal potential (E_f) that is effective in the methane conversion reaction is varied from -0.23 V vs. Standard Hydrogen Electrode (SHE) to 0.7 V vs. SHE, depending on the type of electroactive substances [28]. In BEF1, the applied voltage between the electrodes was 1.0 V, and the electric field in the bulk solution is theoretically 0.3 V/cm. This indicates that the 1st redox peaks were the effective electroactive substances that could contribute to the methane production. It seems that the methane in BEF1 was produced from the acetate by syntrophic metabolisms between the exoelectrogens and the electrotrophic methanogens or by the acetoclastic methanogens.

In BEF2 with an applied voltage of 2 V, the methane production began on the 15th day and reached a maximum value of 43.7 mL/g lignite on the 19th day (Figure 2). Hydrogen production in BEF2 was observed on the 15th day only (Figure 3). The correlation between the hydrogen consumption and methane production was low, similar to BEF1. This indicates that the potential of the hydrogenotrophic methanogenesis in BEF2 was also low. In the cyclic voltammogram for the bulk solution in BEF2, pairs of redox peaks were observed at 0.055/−0.286 V vs. Ag/AgCl (E_f = −0.116 V) and 0.986/0.135 V vs. Ag/AgCl (E_f = +0.561 V). Although the formal potential of the 1st redox peaks shifted in the positive direction somewhat compared to that of BEF1, these peaks may have contributed to the electron transfer for methane production. This suggests that the DIET between the exoelectrogens and the electrotrophs potentially plays a role in the fermentation of the hydrolysis product and the methane production.

Interestingly, the cumulative methane production in BEF1 and BEF2 gradually decreased after increasing to maximum values (Figure 2). It seems that the methane was consumed by methanotrophs. In general, methanotrophs metabolize methane in aerobic conditions and have a unique ability to oxidize a wide range of alkanes, aromatics, and halogenated alkenes [29–31]. However, when the available molecular oxygen is limited, the methanotrophs use sulfate, nitrite, and nitrate as the electron acceptor [31,32]. Recently, it was revealed that the methanotrophs can also use the anode as an electron acceptor to oxidize methane in bioelectrochemical anaerobic systems and can outcompete the methanogens when the available substrate is deficient [32,33]. The anaerobic batch reactors of BEF1 and BEF2 are substrate limited bioelectrochemical reactors with applied voltages, and the hydrolysis product of coal is composed of cyclic compounds. These are conditions under which the methanotrophs could be enriched. It seems that methanotrophs play an important role in the methane conversion of coal, but further studies are needed in the future.

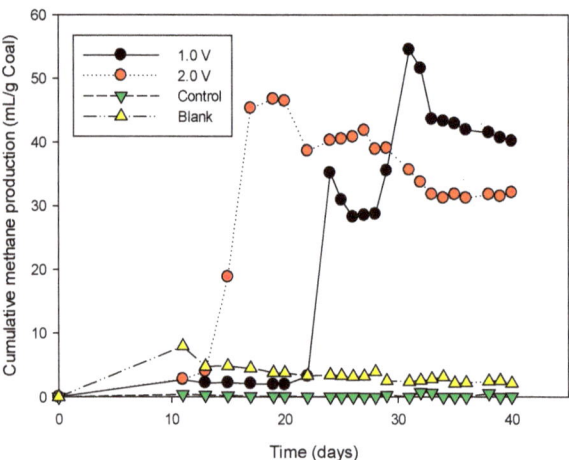

Figure 2. Cumulative methane production in bioelectrochemical anaerobic reactor.

The methane production was very small in the control without an applied voltage and was less than the blank (Figure 2). In the blank, methane was produced from the anaerobic degradation of organic matter contained in the inoculum. It seems that the hydrolysis products of coal had an inhibitory effect on the methane production in the control. Under anaerobic conditions, the hydrolysis products of coal include long chain fatty acids, polycyclic aromatic hydrocarbons, and heterocyclic compounds [2,4]. It is known that the fermentation of the hydrolysis products into precursors, such as acetate and hydrogen, is the rate-limiting step in the entire methane conversion process of coal [2].

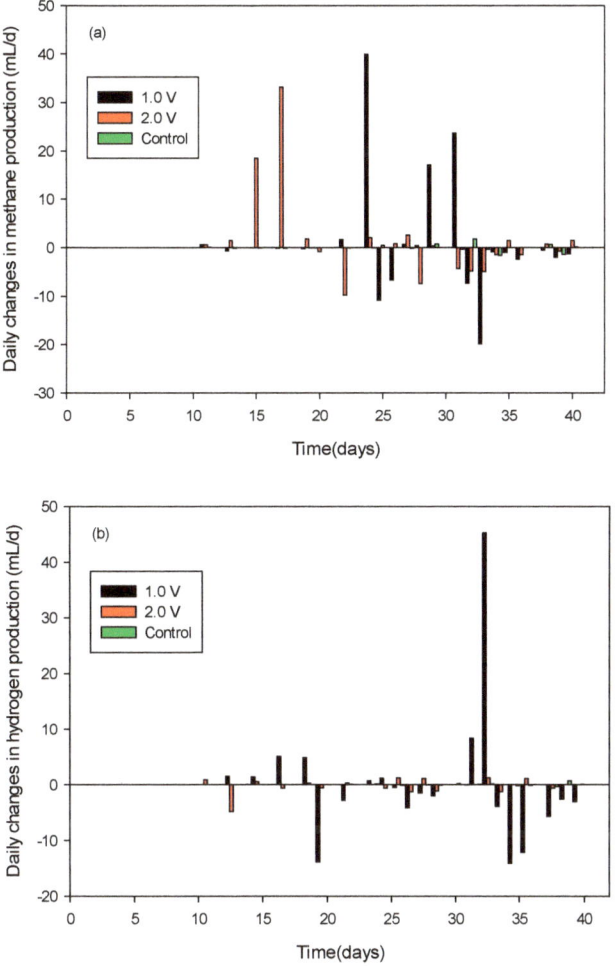

Figure 3. Daily changes in (**a**) methane production and (**b**) hydrogen production.

In previous studies, methane conversion of coal in the anaerobic reactor was observed between 30 and 60 days [3,4]. There is a possibility that the anaerobic microorganisms adapted to the anaerobic reactor for the methane conversion of coal. However, the expected methane production was still less than a few mL/g of coal.

The correlation between hydrogen consumption and methane production in the control was somewhat higher than those in BEF1 and BEF2. This suggests that the methane was produced in the control by indirect interspecies electron transfer via methane precursors, such as acetate or hydrogen. In the bulk solution of the control, redox peaks in the cyclic voltammogram were also observed at 0.08/−0.24 V vs. Ag/AgCl ($E_f = -0.08$ V) and 1.03/0.15 V vs. Ag/AgCl ($E_f = 0.59$ V). However, the small amount of methane production indicates that the redox species in the control did not contribute to the electron transfer for methane production.

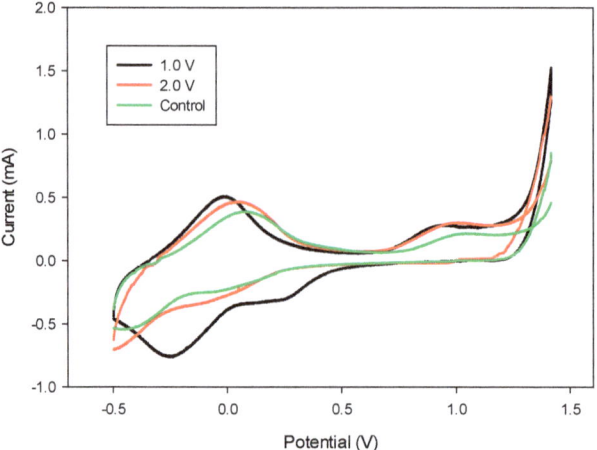

Figure 4. Cyclic voltammogram in the bulk solution for different applied voltages.

3.2. Methane Conversion Potential of Hydrolyzed Product from Coal

The methane conversion of coal experiments was stopped when no further methane production was observed in the reactors. However, the soluble organic matter in BEF1 and BEF2 were still 3.62 and 4.09 g SCOD/L, respectively, and was as high as 4.66 g SCOD/L in the control (Table 1). It is interesting that there was enough organic matter in the anaerobic batch reactors, but the methane production stopped. It is speculated that the residual organic matter was composed of recalcitrant compounds that were difficult to degrade or that the anaerobic microorganisms had lost activity for methane conversion by the toxicity of the compounds. During the anaerobic batch experiments, the pH values were initially 7.36, but increased to 7.79–8.06 (Table 1). In the anaerobic reactors, the accumulation of VFAs reduced the pH, but the alkalinity increased the pH [22,27]. The alkalinities in the anaerobic batch reactors increased from 3824 mg/L as $CaCO_3$ to about 4500 mg/L as $CaCO_3$, but the levels of VFA residuals were very low (Table 1). The alkalinity was generated by the methane production, sulfate reduction, and the ammonia produced from the degradation of nitrogenous compounds [22,27]. In BEF1 and BEF2, the methane production amounts were 116.92–136.27 mL, indicating that the alkalinities were not significantly increased by the methane production. However, the VSSs were reduced from the initial values of 9.25 g/L to 6–8 g/L. This indicates that the alkalinities mainly increased due to the ammonia released from microbial cell lysis due to the limited substrate. In the blank, the residual organic matter was 1.24 g SCOD/L. The hydrolysis products of coal were 2.38–3.42 g SCOD/L in the total organic residue in the anaerobic batch reactors.

In previous studies, the hydrolysis product of coal was composed of several complex cyclic compounds [2,3,13]. These compounds can be converted to the methane precursors by carboxylation, hydroxylation, and methylation for ring opening [8]. The initial value of particulate COD was 11.28 g/L in all anaerobic reactors and decreased to 0.51–1.01 g/L during the methane conversion experiment of coal, indicating that the particulate organic matter containing the coal was at least 91–96% hydrolyzed. However, the VFA residues in BEF1 and BEF2 were only 0.15 g COD/L and 0.19 g COD/L, respectively. This means that the fermentation of the hydrolysis products of coal into the methane precursors is a rate-limiting step in the overall methane conversion of coal [2,4].

In the anaerobic toxicity test, the methane production was severely affected by the content of the coal hydrolysis product in the anaerobic medium (Figure 5a). In the control without the hydrolysis product, the maximum methane production rate was 6.36 mL/day. When the hydrolysis product was added to the anaerobic medium up to 10%, the maximum methane production increased slightly to 6.49 mL/day. However, as the hydrolysis products increased to 20% and 40% in the anaerobic medium, the maximum methane production rates decreased to 2.36 and 1.86 mL/day, respectively. These indicate that the maximum methane production rates were inhibited by 62.9% and 70.8%, respectively. This suggests that the hydrolysis product has a substrate inhibition effect on methane production, and that the maximum methane production rate was 50% inhibited when the hydrolysis product was diluted 5.7-fold (Figure 5b). However, when the hydrolysis product was 10%, the ultimate methane production was 47.4 mL, which was higher than 38.6 mL of the control. The ultimate methane production increased to 52.4 and 49.5 mL with the increase in the hydrolysis products to 20% and 40%, respectively. Although the ultimate methane production depended on the amount of the hydrolysis product added to the anaerobic medium, the increase in the methane production compared to that of the control indicates that the hydrolysis product was metabolized by the anaerobic microorganisms. The methane yield was 175.2 mL/g COD_r in the control, but it decreased from 156.4 mL/g COD_r to 142.3 mL/g COD_r as the hydrolysis product content increased from 10% to 40% (Table 2). This indicates that the hydrolysis products were toxic to the anaerobic microorganisms. The methane yield from the hydrolysis product based on the coal ranged from 106.3 mL/g COD_{rCHP} (55.3 mL/g lignite) to 85.4 mL/g COD_{rCHP} (44.4 mL/g lignite) when the hydrolysis product increased from 10% to 40% in the anaerobic medium, respectively. The dependence of the methane yield on the hydrolysis product content indicates that the hydrolysis product of coal can be converted into methane if it is diluted. However, the substrate inhibition of the hydrolysis product to methane production could be mitigated through additional in-depth studies.

Table 1. Summary of biogenic conversion of coal to methane in bioelectrochemical anaerobic reactor.

Contents		Control	BEF1	BEF2
CH_4 yield (mL/g lignite)		0.75	52.5	43.7
SCOD residual (g/L)		4.66	3.62	4.09
VFAs residual (g COD/L)		0.57	0.18	0.15
Redox peaks in CV	$E_{p,ox}/E_{p,red}$ (V vs. Ag/AgCl)	0.083/−0.241; 1.031/0.154	−0.014/−0.239; 0.913/0.025	0.055/−0.286; 0.986/0.135
	E_f (V vs. Ag/AgCl)	−0.079; +0.593	−0.127; +0.469	−0.116; +0.561
	$I_{p,ox}/I_{p,red}$ (mA)	0.331/0.238; 0.113/0.148	0.473/0.356; 0.217/0.253	0.407/0.244; 0.146/0.011
Alkalinity (mg/L as $CaCO_3$)	Initial	3824	3824	3824
	Final	4026	4182	4462
pH	Initial	7.36	7.36	7.36
	Final	7.79	8.06	7.92
VSS (g/L)	Initial	9.25	9.25	9.25
	Final	5.05	6.10	7.95

Table 2. Summary of the anaerobic toxicity of the hydrolysis product of coal to methane conversion.

Parameter	Control	10%	20%	40%
P_u (mL)	38.6	47.4	52.4	49.5
μ_m (mL/day)	6.36	6.49	2.36	1.86
λ (day)	0	0	0	4.65
Total CH_4 yield (mL/g COD_r)	175.2	156.4	146.2	142.3
CH_4 from CHP (mL/g COD_r CHP)	0.00	106.3	100.0	85.4

(CHP: coal hydrolysis product, g COD).

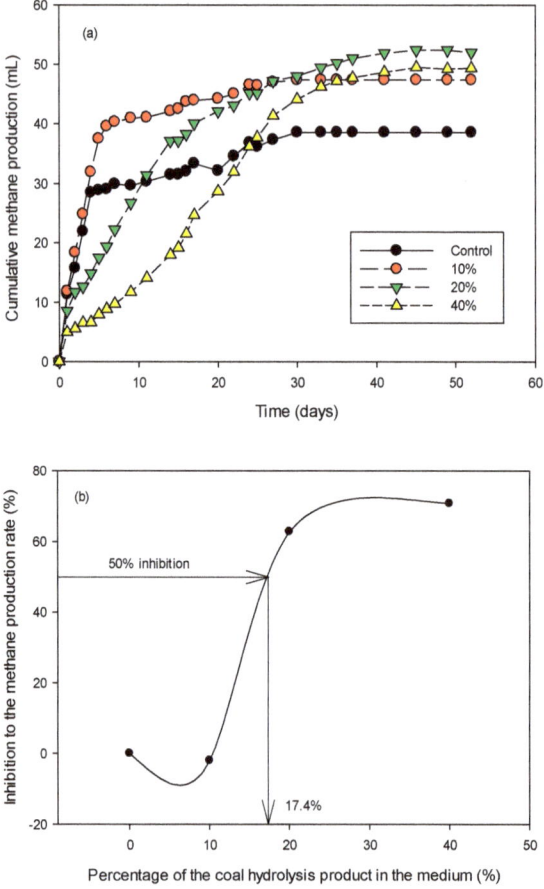

Figure 5. (a) Cumulative methane production and (b) the inhibition to the methane production depending on the amount of the coal hydrolysis product added in the anaerobic toxicity test.

3.3. Implications

Coal is the most buried fossil fuel on the planet and accounts for around 30% of the global energy use [10,11,34]. However, coal releases several pollutants, such as dust, sulfur oxides, nitrogen oxides, and carbon dioxide during use [10,35,36]. Methane is a major component of natural gas, a clean energy resource. The biological conversion of coal to methane is crucial for securing stable energy resources and sustainable development. Therefore, there have been several physicochemical and biological attempts to improve the coal conversion to methane [10,36]. Nevertheless, the methane yield was less than a few mL/g of lignite and required 30 to 85 days [5,37]. The coal conversion process to methane is significantly affected by the physicochemical characteristics of the organic matter contained in the coal. The organic matter in the coal is generally composed of recalcitrant lignified materials that are hydrolyzed into long chain fatty acids, polycyclic aromatic hydrocarbons, and heterocyclic compounds [2,4,5]. In bioelectrochemical anaerobic reactors, the coal conversion to methane was significantly improved by applying 1 V. The methane yield reached 52.5 mL/g lignite (Table 1), which is the highest reported value to the best of our knowledge. However, when the applied voltage increased to 2 V, although the methane yield was slightly reduced to 43.7 mL/g lignite, the methane production from coal was observed after a shorter lag phase time of 13 days. Interestingly, the organic residues of the coal hydrolysis product were still high, with values around 3620–4090 mg

COD/L in the bioelectrochemical anaerobic reactor after methane production. In the anaerobic toxicity tests, the hydrolysis products had produced a substrate inhibition effect for methane production, which was 50% inhibited when the organic hydrolysis product was diluted approximately 5.7 times. However, the organic hydrolysis products were further converted to methane (106.3 mL/g COD_r CHP) by a 10-fold dilution (Table 2). This suggests that the methane production from the hydrolysis product after the bioelectrochemical conversion of coal to methane increases the methane potential of coal to 107.8 mL/g lignite (207.3 mL/g COD in lignite). The methane potential of coal is lower than 350 mL/g COD in glucose, but high enough to be commercially viable. However, additional in-depth studies are necessary to mitigate the inhibition of the coal hydrolysis products on the methane conversion in the field.

4. Conclusions

The polarization of the electrodes in the bioelectrochemical anaerobic reactors greatly improved the coal conversion to methane. The electrode polarized with 1.0 V in the bioelectrochemical anaerobic reactor increased the methane yield of coal to 52.5 mL/g lignite. The electrode polarized with 2 V shortened the time required to produce methane and improved the coal conversion rate. The organic residue of the hydrolysis products had a substrate inhibition effect for methane conversion, and the methane conversion rate was 50% inhibited when the hydrolysis products were diluted 5.7-fold. A 10-fold dilution of the hydrolysis products produced additional methane of 106.3 mL per g COD_r, which amounts to 55.3 mL/g lignite. The total methane potential of coal was improved to 107.8 mL/g lignite by the electrode polarization and the dilution of the hydrolysis products. However, the inhibition of the hydrolysis products to the methane conversion could be mitigated by additional in-depth studies.

Author Contributions: Y.-C.S. and D.H.K. conceived the original idea and Y.-C.S. and D.M.P. designed the study. D.M.P. carried out the experiment and collected the data. Y.-C.S. and D.-H.K. interpreted the data and developed the theory. All authors discussed the data and contributed to the final manuscript.

Acknowledgments: This study was funded by the National Research Foundation of Korea, funded by South Korean Government (NISP), grant number NRF-2017R1E1A1A01075325.

Conflicts of Interest: There is no conflict of interests regarding the publication of this article.

References

1. Colosimo, F.; Thomas, R.; Lloyd, J.R.; Taylor, K.G.; Boothman, C.; Smith, A.D.; Lord, R.; Kalin, R.M. Biogenic methane in shale gas and coal bed methane: A review of current knowledge and gaps. *Int. J. Coal Geol.* **2016**, *165*, 106–120. [CrossRef]
2. Wang, B.; Tai, C.; Wu, L.; Chen, L.; Liu, J.M.; Hu, B.; Song, D. Methane production from lignite through the combined effects of exogenous aerobic and anaerobic microflora. *Int. J. Coal Geol.* **2017**, *173*, 84–93. [CrossRef]
3. Davis, K.J.; Lu, S.; Barnhart, E.P.; Parker, A.E.; Fields, M.W.; Gerlach, R. Type and amount of organic amendments affect enhanced biogenic methane production from coal and microbial community structure. *Fuel* **2018**, *211*, 600–608. [CrossRef]
4. Orem, W.H.; Voytek, M.A.; Jones, E.J.; Lerch, H.E.; Bates, A.L.; Corum, M.D.; Warwick, P.D.; Clark, A.C. Organic intermediates in the anaerobic biodegradation of coal to methane under laboratory conditions. *Org. Geochem.* **2010**, *41*, 997–1000. [CrossRef]
5. Zheng, H.; Chen, T.; Rudolph, V.; Golding, S.D. Biogenic methane production from Bowen Basin coal waste materials. *Int. J. Coal Geol.* **2017**, *169*, 22–27. [CrossRef]
6. Shin, H.S.; Song, Y.C. A model for evaluation of anaerobic degradation characteristics of organic waste: Focusing on kinetics, rate-limiting step. *Environ. Technol.* **1995**, *16*, 775–784. [CrossRef]
7. Song, Y.C.; Feng, Q.; Ahn, Y. Performance of the Bio-electrochemical Anaerobic Digestion of Sewage Sludge at Different Hydraulic Retention Times. *Energy Fuels* **2016**, *30*, 352–359. [CrossRef]

8. Nzila, A. Biodegradation of high-molecular-weight polycyclic aromatic hydrocarbons under anaerobic conditions: Overview of studies, proposed pathways and future perspectives. *Environ. Pollut.* **2018**, *239*, 788–802. [CrossRef] [PubMed]
9. Sharma, A.; Singh, S.B.; Sharma, R.; Chaudhary, P.; Pandey, A.K.; Ansari, R.; Vasudevan, V.; Arora, A.; Singh, S.; Saha, S.; et al. Enhanced biodegradation of PAHs by microbial consortium with different amendment and their fate in in-situ condition. *J. Environ. Manag.* **2016**, *181*, 728–736. [CrossRef] [PubMed]
10. Zhang, J.; Liang, Y.; Harpalani, S. Optimization of methane production from bituminous coal through Biogasification. *Appl. Energy* **2016**, *183*, 31–42. [CrossRef]
11. Fuertez, J.; Nguyen, V.; McLennan, J.D.; Adams, D.J.; Han, K.B.; Sparks, T.D. Optimization of biogenic methane production from coal. *Int. J. Coal Geol.* **2017**, *183*, 14–24. [CrossRef]
12. Jones, E.J.P.; Voytek, M.A.; Corum, M.D.; Orem, W.H. Stimulation of Methane Generation from Nonproductive Coal by Addition of Nutrients or a Microbial Consortium. *Appl. Environ. Microbiol.* **2010**, *76*, 7013–7022. [CrossRef] [PubMed]
13. Ghosh, S.; Jha, P.; Vidyarthi, A.S. Unraveling the microbial interactions in coal organic fermentation for generation of methane—A classical to metagenomic approach. *Int. J. Coal Geol.* **2014**, *125*, 36–44. [CrossRef]
14. Bao, Y.; Huang, H.; He, D.; Ju, Y.; Qi, Y. Microbial enhancing coal-bed methane generation potential, constraints and mechanism. *J. Nat. Gas Sci. Eng.* **2016**, *35*, 68–78. [CrossRef]
15. Feng, Q.; Song, Y.C.; Ahn, Y. Electroactive microorganisms in bulk solution contribute significantly to methane production in bioelectrochemical anaerobic reactor. *Bioresour. Technol.* **2018**, *259*, 119–127. [CrossRef] [PubMed]
16. Doyle, L.E.; Marsili, E. Methods for enrichment of novel electrochemically-active microorganisms. *Bioresour. Technol.* **2015**, *195*, 273–282. [CrossRef] [PubMed]
17. Zhao, Z.; Zhang, Y.; Quan, X.; Zhao, H. Evaluation on direct interspecies electron transfer in anaerobic sludge digestion of microbial electrolysis cell. *Bioresour. Technol.* **2016**, *200*, 235–244. [CrossRef] [PubMed]
18. Feng, Q.; Song, Y.C.; Yoo, K.; Kuppanan, N.; Subudhi, S.; Lal, B. Polarized electrode enhances biological direct interspecies electron transfer for methane production in upflow anaerobic bioelectrochemical reactor. *Chemosphere* **2018**, *204*, 186–192. [CrossRef] [PubMed]
19. Rotaru, A.E.; Shrestha, P.M.; Liu, F.; Shrestha, M.; Shrestha, D.; Embree, M.; Zengler, K.; Wardman, C.; Nevin, K.P.; Lovley, D.R. A new model for electron flow during anaerobic digestion: Direct interspecies electron transfer to Methanosaeta for the reduction of carbon dioxide to methane. *Energy Environ. Sci.* **2014**, *7*, 408–415. [CrossRef]
20. Shrestha, P.M.; Rotaru, A.E. Plugging in or going wireless: Strategies for interspecies electron transfer. *Front. Microbiol.* **2014**, *5*, 237–244. [CrossRef] [PubMed]
21. Song, Y.C.; Kwon, S.J.; Woo, J.H. Mesophilic and thermophilic temperature co-phase anaerobic digestion compared with single-stage mesophilic- and thermophilic digestion of sewage sludge. *Water Res.* **2004**, *38*, 1653–1662. [CrossRef] [PubMed]
22. Feng, Q.; Song, Y.C.; Bae, B.U. Influence of applied voltage on the performance of bioelectrochemical anaerobic digestion of sewage sludge and planktonic microbial communities at ambient temperature. *Bioresour. Technol.* **2016**, *220*, 500–508. [CrossRef] [PubMed]
23. Woo, J.H.; Song, Y.C. Influence of temperature and duration of heat treatment used for anaerobic seed sludge on biohydrogen fermentation. *KSCE J. Civ. Eng.* **2010**, *14*, 141–147. [CrossRef]
24. Modestra, J.A.; Navaneeth, B.; Mohan, S.V. Bio-electrocatalytic reduction of CO_2: Enrichment of homoacetogens and pH optimization towards enhancement of carboxylic acids biosynthesis. *J. CO_2 Util.* **2015**, *10*, 78–87. [CrossRef]
25. Kadier, A.; Kalil, M.S.; Chandrasekhar, K.; Mohanakrishna, G.; Saratale, G.D.; Saratale, R.G.; Kumar, G.; Pugazhendhi, A.; Sivagurunathan, P. Surpassing the current limitations of high purity H_2 production in microbial electrolysis cell (MECs): Strategies for inhibiting growth of methanogens. *Bioelectrochemistry* **2018**, *119*, 211–219. [CrossRef] [PubMed]
26. Wang, L.; Singh, L.; Liu, H. Revealing the impact of hydrogen productionconsumption loop against efficient hydrogen recovery in single chamber microbial electrolysis cells (MECs). *Int. J. Hydrogen Energy* **2018**, *43*, 13064–13071. [CrossRef]

27. Feng, Q.; Song, Y.C.; Yoo, K.; Kuppanan, N.; Subudhi, S.; Lal, B. Bioelectrochemical enhancement of direct interspecies electron transfer in upflow anaerobic reactor with effluent recirculation for acidic distillery wastewater. *Bioresour. Technol.* **2017**, *241*, 171–180. [CrossRef] [PubMed]
28. Lim, S.S.; Yu, E.H.; Daud, W.R.W.; Kim, B.H.; Scott, K. Bioanode as a limiting factor to biocathode performance in microbial electrolysis cells. *Bioresour. Technol.* **2017**, *238*, 313–324. [CrossRef] [PubMed]
29. Murell, J.C.; Radajewski, S. Cultivation-independent techniques for studying methanotroph ecology. *Res. Microbiol.* **2000**, *151*, 807–814. [CrossRef]
30. Pandey, A.; Mai, V.T.; Vu, D.Q.; Bui, T.P.L.; Mai, T.L.A.; Jensen, L.S.; Neergaard, A. Organic matter and water management strategies to reduce methane and nitrous oxide emissions from rice paddies in Vietnam. *Agric. Ecosyst. Environ.* **2014**, *196*, 137–146. [CrossRef]
31. Wang, L.; Pan, Z.; Xu, H.; Wang, C.; Gao, L.; Zhao, P.; Dong, Z.; Zhang, J.; Cui, G.; Wang, S.; et al. The influence of nitrogen fertiliser rate and crop rotation on soilmethane flux in rain-fed potato fields inWuchuan County, China. *Sci. Total Environ.* **2015**, *537*, 93–99. [CrossRef] [PubMed]
32. Gao, Y.; Ryu, H.; Rittmann, B.E.; Hussain, A.; Lee, H.S. Quantification of the methane concentration using anaerobic oxidation of methane coupled to extracellular electron transfer. *Bioresour. Techno.* **2017**, *241*, 979–984. [CrossRef] [PubMed]
33. Liu, S.; Feng, X.; Li, X. Bioelectrochemical approach for control of methane emission from wetlands. *Bioresour. Tech.* **2017**, *241*, 812–820. [CrossRef] [PubMed]
34. Huang, Z.; Liers, C.; Ullrich, R.; Hofrichter, M.; Urynowicz, M.A. Depolymerization and solubilization of chemically pretreated powder river basin subbituminous coal by manganese peroxidase (MnP) from Bjerkandera adusta. *Fuel* **2013**, *112*, 295–301. [CrossRef]
35. Ryu, H.W.; Chang, Y.K.; Kim, S.D. Microbial coal desulfurization in an airlift bioreactor by sulfur-oxidizing bacterium thiobacillus ferrooxidans. *Fuel Proc. Technol.* **1993**, *36*, 267–275. [CrossRef]
36. Yi, L.; Feng, J.; Qin, Y.H.; Li, W.Y. Prediction of elemental composition of coal using proximate analysis. *Fuel* **2017**, *193*, 315–321. [CrossRef]
37. Xiao, D.; Peng, S.P.; Wang, B.Y.; Yan, X.X. Anthracite bio-degradation by methanogenic consortia in Qinshui basin. *Int. J. Coal Geol.* **2013**, *116*, 46–52. [CrossRef]

© 2018 by the authors. Licensee MDPI, Basel, Switzerland. This article is an open access article distributed under the terms and conditions of the Creative Commons Attribution (CC BY) license (http://creativecommons.org/licenses/by/4.0/).

Article

Separation of Acetate Produced from C1 Gas Fermentation Using an Electrodialysis-Based Bioelectrochemical System

Jiyun Baek, Changman Kim, Young Eun Song, Hyeon Sung Im, Mutyala Sakuntala and Jung Rae Kim *

School of Chemical and Biomolecular Engineering, Pusan National University, 63 Busandeahak-ro, Geumjeong-Gu, Busan 46241, Korea; bjyjupiter@naver.com (J.B.); aslongas77@naver.com (C.K.); duddms37@naver.com (Y.E.S.); gj9338@naver.com (H.S.I.); sakuntala1819@gmail.com (M.S.)
* Correspondence: j.kim@pusan.ac.kr; Tel: +82-1-510-2393; Fax: +82-51-510-3943

Received: 1 September 2018; Accepted: 7 October 2018; Published: 16 October 2018

Abstract: The conversion of C1 gas feedstock, such as carbon monoxide (CO), to useful platform chemicals has attracted considerable interest in industrial biotechnology. One conversion method is electrode-based electron transfer to microorganisms using bioelectrochemical systems (BESs). In this BES system, acetate is the predominant component of various volatile fatty acids (VFAs). To appropriately separate and concentrate the acetate produced, a BES-type electrodialysis cell with an anion exchange membrane was constructed and evaluated under various operational conditions, such as applied external current, acetate concentration, and pH. A high acetate flux of 23.9 mmol/m^2·h was observed under a −15 mA current in an electrodialysis-based bioelectrochemical system. In addition, the initial acetate concentration affected the separation efficiency and transportation rate. The maximum flux appeared at 48.6 mmol/m^2·h when the acetate concentration was 100 mM, whereas the effects of the initial pH of the anolyte were negligible. The acetate flux was 14.9 mmol/m^2·h when actual fermentation broth from BES-based CO fermentation was used as a catholyte. A comparison of the synthetic broth with the actual fermentation broth suggests that unknown substances and metabolites produced from the previous bioconversion process interfere with electrodialysis. These results provide information on the optimal conditions for the separation of VFAs produced by C1 gas fermentation through electrodialysis and a combination of a BES and electrodialysis.

Keywords: electrodialysis; bioelectrochemical system; microbial fuel cell; C1 gas; carbon monoxide; acetate

1. Introduction

The biological conversion of industrial waste gases containing carbon dioxide and carbon monoxide are being highlighted to reduce the emissions of greenhouse gases and simultaneously produce the building blocks of fuel and more useful commodity chemicals [1,2]. Among them, CO, which is toxic and recalcitrant to the environment, accounts for 50 to 70% of the effluent gas from steel factories. Hence, appropriate treatment technologies are anticipated. Recently, Im et al. (2018) reported that a bioelectrochemical system (BES) could compensate for the limitation of natural biological CO conversion and enhance the production of volatile fatty acids [3]. The applied potential of the BES supplies reducing power for autotrophic microorganisms and improves the yield of C1 gas conversion and cell growth [4–8].

The metabolites produced from BES-based C1 fermentation may contain acetate as well as various volatile fatty acids (VFAs) and alcohols [9,10]. Therefore, additional separation processes are needed to isolate and/or concentrate acetate from fermentation broth. In a conventional study, the separation of ionic metabolites was carried out using chemical and physical methods, such as acidification, ion-exchange, crystallization, and adsorption. On the other hand, these attempts may need to be moderated with the recent trends of environmentally and economically sustainable research and development [11,12]. For example, in the case of acidification or ion exchange, considerable amounts of acid and alkali are consumed during the operation, which is problematic. Regarding crystallization or adsorption, additional purification and chemical waste discharge have been a concern.

Electrodialysis (ED) is a technology to separate and enrich target substances by transferring the ionic forms through a selectively transmissible ion exchange membrane under an electrochemically induced oxidation/reduction reaction [11]. The separation of fermentation metabolites by electrodialysis was proposed to prevent the inhibition of lactic acid [13]. Im et al. reported that BES-based C1 gas conversion produced up to 8.4 g/L of acetate, which is an industrially useful intermediate chemical and a source for further biosynthesis. Recently, many research groups have attempted to separate acetate from a range of wastewater or microbial fermentation broths [14,15]. In a normal fermentation broth, the anionic form of acetate is the dominant species rather than acetic acid because the culture condition is generally near neutral pH. Therefore, acetate species can be separated using the ion exchange membrane of an electrodialysis cell. In electrodialysis, the H^+/OH^- ion can be supplied continuously by electrochemical control in electrodialysis, and this can provide a driving force to separate various metabolites from the fermentation broth without the need for additional chemical reagents, such as salts. Moreover, it is capable of separating and concentrating high purity substances efficiently compared to other methods, enabling applications in a wide range of industrial processes, including food and biofuel production [16,17]. In particular, there have been many applications of electrodialysis in bioelectrochemical systems [18–20]. For example, ethyl acetate was produced through biphasic esterification, and acetic acid was separated from the fermentation broth by electrodialysis [21]. In addition, acetic acid, which was produced from carbon dioxide in a three-chamber bioelectrochemical system, was separated by electrodialysis with yields of up to 13.5 g/L over a 43 day period [5].

The system configuration of BES and electrodialysis have some similarity in terms of using an ion exchange membrane (or separator) and electrical input (or output) to (or from) the reactor. Thus, electrodialysis allows the direct production and isolation of the target metabolites from C1 gas fermentation. On the other hand, the most important and problematic issue of separation by electrodialysis is membrane fouling [22]. In the sludge, wastewater or fermentation broth, there are not only secondary metabolic products, but also unused growth media components and a large number of cells [23]. These undesirable substances or microbial cells attach to the surface of the membrane and/or block the functional group of the ion exchange membrane during the electrodialysis process, eventually resulting in a decrease in separation efficiency [24]. To solve these problems, pretreating the fermentation broth before introduction to electrodialysis or various modification methods of the ion exchange membrane have been suggested [23,25,26].

This study examined the operational parameters in electrodialysis to separate acetate, which is applicable to C1 gas fermentation (Figure 1). The optimal conditions in the synthetic broth were investigated and applied to the fermentation broth. The efficiency and flux of acetate separation were compared in an actual fermentation broth and synthetic solution. The aim of this study was to assess the potential of a combination of electrodialysis with BES-based C1 gas fermentation.

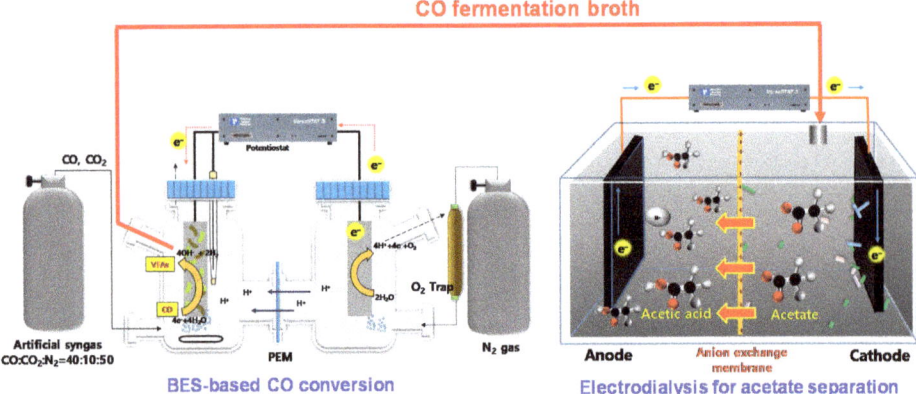

Figure 1. Separation and concentration of acetate from CO fermentation broth by electrodialysis cell. BES: Bioelectrochemical system.

2. Materials and Methods

2.1. Configuration of Bioelectrochemical System and Electrodialysis Reactor

An H-type BES reactor was used, as described previously [3,4]. Two media bottles (310 mL capacity) were joined with a glass tube and a proton exchange membrane (Nafion 117; Dupont, Wilmington, DE, USA) and held with a clamp [27]. A 4 cm × 5 cm piece of graphite felt (Cera Materials, Port Jervis, NY, USA) was implemented as the cathode electrode. Graphite granules (40 g) were added to the anode chamber, and 5 cm of a graphite rod connected with titanium wire was used as the current collector from the graphite granules. The cathode potential (−1.1 V vs. Ag/AgCl) was applied continuously through a multi-channel potentiostat (WMPG1000K8, Won-A tech, Seoul, Korea) during the experiment to support the BES-based biological CO conversion. A feed gas (N_2:CO:CO_2 = 50:40:10) was continuously provided into the cathode chamber at a flow rate of 10 mL/min. All experimental conditions performed were in accordance with the research reported by Im et al. [3].

The electrodialysis (ED) reactor used in this experiment consisted of an acrylic anode and cathode chamber; each chamber had a working volume of 73.5 mL (7 × 7 × 1.5 cm^3) (See Figure S1). Both electrodes were made of carbon paper (surface area of 42.25 cm^2, 120-TGP-H-120, Toray, Japan), and connected to a circuit via a carbon fiber (20 cm). An anion exchange membrane (49 cm^2, FKB-PK-130, Fumasep, Bietigheim-Bissingen, Germany) was used as the ion exchange membrane for the cell, and it was rinsed with a 0.5 M NaCl solution for 24 h prior to use. A potentiostat (WMPG1000, WonA Tech, Korea) in galvanostatic mode was used to apply a current to the reactor. To examine acetate separation from a realistic fermentation broth, both BES and ED were connected, as shown in Figure 1. In some cases, centrifuged fermentation broth, as described in Section 2.2, was introduced into the electrodialysis cell to examine the effects of particulates and cells in the media.

2.2. Composition of Electrolyte

Two types of catholytes were used to examine acetate transportation across the ion exchange membrane: synthetic broth and fermentation broth. The synthetic broth contained a CO/CO_2 fermentation medium, which was composed of the following (per liter): 1.5 g KH_2PO_4, 2.9 g K_2HPO_4, 2.0 g $NaHCO_3$, 0.5 g NH_4Cl, 0.09 g $MgCl \cdot 6H_2O$, 0.0225 g $CaCl_2 \cdot 2H_2O$, and 0.5 g yeast extract. Sodium acetate (20 mM to 100 mM for each reaction condition) was added to the catholyte to examine transportation through the membrane. The fermentation medium was made by slightly modifying the synthetic broth by also adding 2.11 g of sodium-2 bromoethanesulfonate as a methanogene inhibitor, 2 ml of Pfennig's trace element solution, and 5 mL of a vitamin solution [3]. The pH was adjusted to 6.0 with 1 M H_2SO_4 and 1 M NaOH. In some experiments, centrifugation was conducted at 7500 RPM and 15 min to remove the cells and precipitates produced from former fermentation. Streptomycin (20 µg/mL) was added as an antibiotic to prevent acetate consumption due to contamination. The anode electrolyte consisted of the following ingredients (per liter): 0.8 g K_2HPO_4, 1.0 g NH_4Cl, 2.0 g KCl, 0.15 g $CaCl_2·2H_2O$, 2.4 g $MgCl·6H_2O$, 4.8 g NaCl, and 10.08 g $NaHCO_3$ [5].

2.3. Operation of Electrodialysis Reactor

The cathode and anode electrodes were set as the working and counter electrodes, respectively. The current applied to the cathode ranged from 0 to −15 mA using a galvanostatic method. The electrodialysis cells were located in the incubator at 25 ± 1 °C and gently shaken at 30 rpm.

2.4. Analyses

A liquid sample (<300 µL) was taken from each chamber periodically. The liquid samples were filtered through a 0.2 µm syringe filter, acidified by HCl to prevent acetate volatilization, and stored in a freezer at −80 °C. The samples were analyzed by gas chromatography (GC, 7890B, Agilent Technologies, Santa Clara, CA, USA) and high-performance liquid chromatography (HPLC, HP 1100 series, Agilent Technologies, Santa Clara, CA, USA). The experiment was conducted in duplicate, and the analyses were carried out in duplicate. The initial and final pH were measured using a pH meter (Orion 420A+, Thermo Orion, USA). The current efficiency (η_A) was estimated using the following equation:

$$\eta_A = \frac{\Delta N_A}{iA\Delta t/F} \qquad (1)$$

where ΔN_A is the change in the molarity of acetate, i is the current density, A is the membrane area, F is the faraday constant (96485 C/mol = 26.8 Ah/mol), and Δt is the interval of time [28].

The flux (J_A) of acetate from the cathode to anode chamber was calculated using the following equation:

$$J_A = \frac{\Delta m A}{A \Delta t} \qquad (2)$$

where Δm is the amount of acetate transported from the cathode to the anode chamber, A is the membrane area, and Δt is the interval of time.

3. Results and Discussion

3.1. Different Applied Current on Acetate Transportation in BES

Acetate transport across the ion exchange membrane is affected by the applied potential and current in microbial fuel cells [5,11,29,30]. Therefore, the changes in acetate concentration in both the anode and cathode chambers were examined while various currents (−5 to −15 mA) were applied to the cell (Figure 2). In the absence of an applied current, the final acetate concentration of 9.17 mM was transported to the anode chamber during 16 h of operation, indicating that acetate had diffused to the anode due to the concentration gradient. On the other hand, acetate transport was increased to 12.55 mM when a current was applied across the electrodes (−5 mA). Under −15 mA application, 24.98 mM of acetate was transported to the anode chamber. An externally applied current can drive the electrochemical reaction and actively move acetate anions against the concentration gradient between the anode and cathode chambers (Figure 2B–D) over 16 h, whereas the acetate only diffused naturally in the control (i.e., under the absence of an applied current) (Figure 2A).

The amount of acetate transportation increased with increasing current in BES. On the other hand, the estimated current efficiency on the applied potential decreased at a higher current (Table 1). The current efficiency estimated by Equation (1) was higher (54.4 ± 0.2%) under a lower applied current (−5 mA), whereas it decreased at a higher current (36.1 ± 1.2% at −15 mA) (Table 1). On the other hand, the acetate flux across the membrane was 23.9 ± 0.8 mmol/m²·h at −15 mA, whereas it decreased at a lower applied current (8.8 ± 0.4 mmol/m²·h at −5 mA) (Table 1). The driving force for acetate anion transportation by electrodialysis is lost under a higher current in electrodialysis. These results are consistent with the previous observation that the selectivity for ions at a low current density was higher than that at a high current density [31]. At a high current density, the current efficiency was reduced because the driving force was dissipated by the movement of other ions in addition to the target acetate, and resistance in the ion exchange membrane.

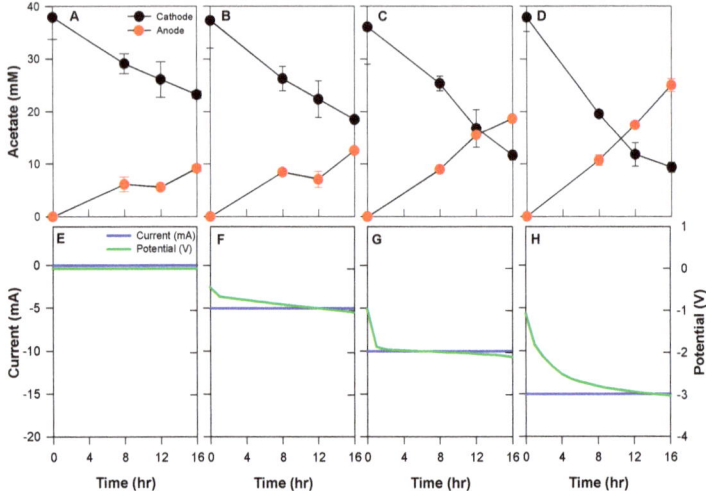

Figure 2. Acetate transfer from the cathode to the anode chamber under different current conditions. Without current application (**A,E**); −5 mA (**B,F**); −10 mA (**C,G**); −15 mA (**D,H**) during 16 h of operation.

Table 1. Entire migration amount of acetate in the cathode chamber, total applied current, current efficiency, and acetate flux.

	Conditions	Acetate Migration from the Cathode (mM)	Total Applied Current (C)	Current Efficiency (%)	Acetate Flux (mmol/m²·h)
Applied current	0 mA	14.7 ± 4.9	-	-	8.8 ± 0.4
	−5 mA	18.8 ± 5.2	288.0	54.4 ± 0.2	12.0 ± 0.0
	−10 mA	24.4 ± 7.9	576.0	40.4 ± 0.6	17.8 ± 0.3
	−15 mA	28.5 ± 3.6	864.0	36.1 ± 1.2	23.9 ± 0.8
Acetate concentration	20 mM	16.3 ± 2.1	864.0	23.8 ± 0.6	15.8 ± 0.4
	40 mM	30.1 ± 4.5		40.2 ± 1.2	26.6 ± 0.8
	80 mM	44.2 ± 9.3		63.1 ± 2.6	41.8 ± 1.7
	100 mM	55.9 ± 11.3		73.4 ± 4.6	48.6 ± 3.0
pH test	2.0	31.2 ± 3.1	864.0	43.4 ± 2.8	28.2 ± 1.8
	4.0	30.6 ± 6.3		42.8 ± 3.4	28.4 ± 2.2
	6.0	28.9 ± 3.9		40.1 ± 4.1	26.6 ± 2.7
Catholyte composition	Synthetic	25.3 ± 0.9	864.0	34.5 ± 1.2	22.9 ± 0.8
	Fermented with centrifuge	11.4 ± 1.2		22.5 ± 1.5	14.9 ± 1.0
	Fermented without centrifuge	11.1 ± 0.3		18.6 ± 0.7	12.3 ± 0.4

3.2. Effect of Different Acetate Concentration

The effects of the initial acetate concentration (20 to 100 mM) on electrodialysis were investigated at an applied current of −15 mA (Figure 3). The acetate flux was estimated to be 48.6 ± 3.0 mmol/m²·h at an initial acetate concentration of 100 mM, whereas it decreased proportionally to 15.8 ± 0.4 mmol/m²·h with 20 mM (Figure 3 and Table 1). At the highest acetate concentration (100 mM), the current efficiency (73.4%) was much higher than that at a lower concentration (20 mM vs. 23.8%) (Table 1). These results suggest that a higher efficiency of acetate separation can be obtained at a higher acetate concentration. When no current was applied to the cell, separation was carried out by diffusion depending on the acetate concentration. This indicates that, in addition to the applied current, diffusion plays an important role in the transport of acetate [29]. Accordingly, a higher acetate concentration is required for optimal process efficiency, even though the performance of electrodialysis is also related to the reactor configuration. Im et al. examined the fermentation of acetate production from CO by electrosynthesis and revealed the productivity of acetate at a maximum of 8.4 g L^{-1} in a BES [3]. Therefore, electrodialysis-driven acetate separation around the maximum was examined in the electrodialysis cell. The separation of acetate at this point is expected to increase both the growth of microorganisms and the acetate productivity from CO conversion.

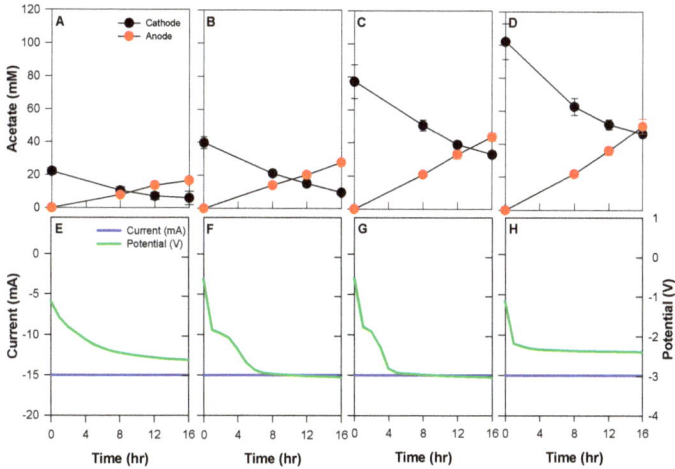

Figure 3. Change in acetate concentration by electrodialysis under various initial acetate concentrations in the cathode chamber, 20 mM (**A,E**), 40 mM (**B,F**), 80 mM (**C,G**), and 100 mM (**D,H**). The applied potential was fixed to −15 mA.

3.3. Effect of Different Initial Anodic pH

The pH of the anode chamber can also be an important factor for the efficient separation of acetate. The pK$_a$ of acetate is 4.76. Hence, acetate exists mainly as an ionized form in the catholyte in the cathode chamber (~pH 6.0), which is provided from the former BES reactor. To examine the effects of the anodic pH in electrodialysis, the anodic pH was adjusted from 2.0 to 6.0 while the cathodic pH was fixed to 6.0 because the pH from the effluent from the former C1 gas fermentation is approximately 6.0. As shown in Figure 4, the pH effect on acetate separation was negligible, and the current efficiencies were estimated to be approximately 40–43% under these pH conditions (Table 1). In electrodialysis cells, the following oxidation and reduction reactions take place in the anode (3) and cathode chamber (4);

$$H_2O \rightarrow 2H^+ + 1/2O_2 + 2e^- \tag{3}$$

$$2H_2O + 2e^- \rightarrow 2OH^- + H_2 \tag{4}$$

The H$^+$ produced by water electrolysis reaction (3) reduces the pH in the anode chamber continuously, which eventually approaches pH 2.0, even if the initial pH of the anode chamber is higher than pH 2.0. The final pH of the anode chamber in the tested reactors was pH 1.9 to 2.0, which converged from a varied initial pH 2.0 to 6.0. This suggests that proton transport from the anode to the cathode in the reverse direction of acetate anion species separation might be limited by the ion exchange membrane [32].

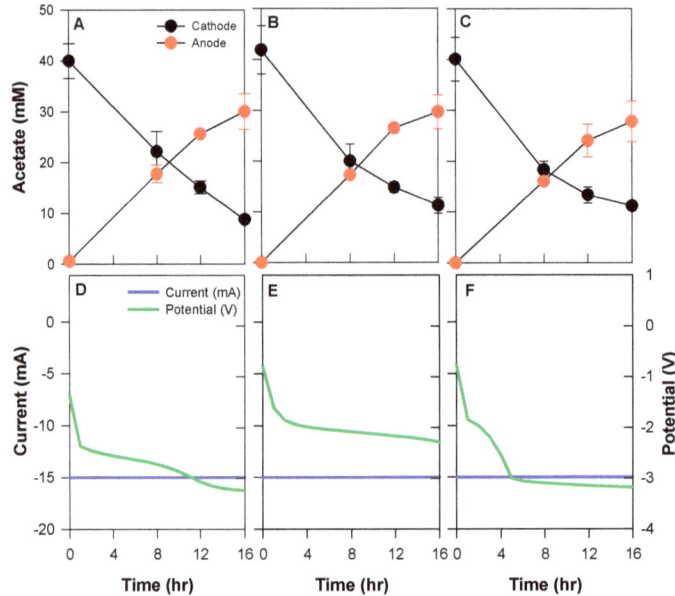

Figure 4. Changes in the acetate concentration by electrodialysis when the initial pH of the anode was varied. Initial anode pH 2.0 (**A,D**); pH 4.0 (**B,E**); pH 6.0 (**C,F**).

3.4. Acetate Separation from the Actual Fermentation Broth

The combination of acetate fermentation followed by electrodialysis-based acetate separation has been highlighted for biological C1 gas conversion. Based on the results of the above experiments, synthetic and fermentation broths containing acetate were compared for the separation of acetate in the electrodialysis cell. First, the cell and precipitate in the former fermentation broth were removed by centrifugation to exclude the effects of particulates in the broth. The final acetate concentrations with the fermentation and synthetic broth were 15.6 mM and 23.9 mM, respectively (Figure 5). The results show that approximately 20% less acetate in the fermentation broth (i.e., effluent from the former electrosynthesis process) is transported to the anode chamber than the synthetic solution, even when the particulates were removed by centrifugation (Figure 5B). A similar but slightly lower acetate separation was obtained using a non-centrifuged fermentation broth (i.e., realistic cultivation broth from BES) (13.30 mM) during the 16 h of operation (Figure 5C). The other metabolites from C1 gas fermentation in BES, such as butyrate, propionate and iso-butyrate [3], hinder acetate separation significantly. As observed in the GC analysis results, unlike the synthetic medium, the fermentation broth contains various volatile fatty acids as well as acetate (Figure S2C,D). Among these metabolites, the longer chain VFAs, such as propionate, may pass through the membrane competitively with acetate, which might reduce the rate and efficiency of acetate separation. The GC results also show clearly that propionate has passed through the anion exchange membrane used in this study (Figure S2A,B). The competitive flux of these other anions is considered to be the cause of the relatively low current efficiency for acetate separation [21]. Because the former BES process was usually inoculated with inoculum, including sludge and isolated microorganisms, it contained a variety of particulates, colloidal and dissolved fractions, all of which can act as inhibitors and potential foulants. Microorganisms and soluble substances potentially cause membrane fouling, which decreases the electrodialysis performance [24,33,34]. Ghasemi et al. reported that the microorganisms attached to the membrane surface and the biofilm formation are major factors reducing the separation efficiency in the electrodialysis cell [22]. After the operation of the electrodialysis cell with a non-centrifuged fermentation broth, contamination by unknown substances was observed on the cathodic electrode, which was different from the synthetic and centrifuged fermentation broth (Figure S3D–F). These contaminants on the electrode and membrane may lower the current efficiency of acetate separation from the non-centrifuged fermentation broth in the electrodialysis cell.

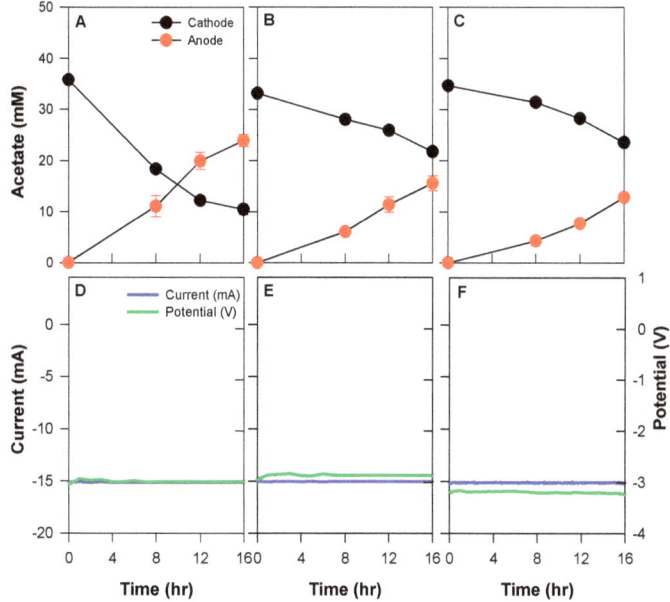

Figure 5. Comparison of acetate transportation in an electrodialysis cell with synthetic media containing acetate (**A,D**), and centrifuged C1 gas fermentation broth (**B,E**) and non-centrifuged C1 gas fermentation broth (**C,F**).

4. Implication and Conclusions

Acetate is one of the primary metabolites from C1 gas fermentation and is a useful intermediate chemical for further biosynthesis [35]. On the other hand, this fermentation broth contains a variety of components, such as ion species, VFAs, and microorganisms as well as acetate. Therefore, the development of appropriate acetate separation is essential, which accounts for between 30% and 40% of the total process cost [36]. In this respect, the production of acetate through BES and the separation of the produced acetate by electrodialysis may provide an appropriate platform for the in-situ processing and separation of acetate. Through these experiments, it was confirmed that a maximum acetate flux of 48.6 ± 3.0 mmol/m^2·h can be achieved using the synthetic broth. When the C1 fermentation broth was applied, the flux was 14.9 ± 1.0 mmol/m^2·h, which was approximately half that of the synthetic broth (22.9 ± 0.8 mmol/m^2·h) under the same conditions, probably due to the various substances and other longer chain VFAs in the fermentation broth. The obtained separation rate and efficiency are comparable to the previous results of electrodialysis (Table 2).

There are still challenges that need to be overcome before this system can be applied to an actual industrial environment; several studies to solve these problems are underway. To solve the fouling of an ion-exchange membrane, some research groups have focused on modifying the membrane by a treatment with polymer compounds, such as poly (sodium 4-styrene sulfonate) (PSS)/poly(diallyldimethylammonium chloride) (PDADMAC) [34]. An ultra-low voltage customized DC–DC booster circuit [37] and maximum power point tracking (MPPT) [38] may provide an affordable voltage and current for self-sustained electrodialysis applications. Although further studies will be needed in the future, these results may provide a basis for techniques to isolate acetate from actual fermentation end products and culture broth from bioelectrochemical systems.

Table 2. Volatile fatty acid separation using various ion exchange membranes and resins.

Volatile Fatty Acids	Separated Acetate Titer	Separation Material & Membrane	Applied Potential or Current	Current Efficiency (%)	Reference
Acetate	1.21 mol/m^3 during 32 days	Anion exchange membrane (1.77 cm^2, AMI-7001)	−800 mV (vs. SHE)	-	[39]
Acetate	0.379 kg/m^2·d	Anion exchange membrane (64 cm^2, AM-7001)	20 Am^{-2} (vs. Ag/AgCl)	36% (coulombic efficiency)	[21]
Acetate	225 mM during 43 days	Anion exchange membrane (Fumatech FAB)	−50 mA (vs. SHE)	-	[5]
Acetate	100 mg/g (acetate sorption)	Anion exchange resin (35 g, Amberlite TM FPA53)	-	-	[40]
Succinic acid	15.7 g/dm^3 during 180 min	EDMB stack: 10 Bipolar (PC 200bip), 10 Anion exchange (PC 200D), Cation exchange(PC-SK), (207 cm^2 each)	120 A/m^2	14.3%	[41]
Lactic acid	1.46 mol/L during 15 h	EDMB stack: Bipolar (40 cm^2, BMP-1), Anion exchange (40 cm^2, FAS-PET-130), Cation exchange (40 cm^2, JCM-II-05)	40 mA/cm^2	-	[42]
Acetate	12.3 ± 0.4 mmol/m^2·h	Anion exchange membrane (49 cm^2, FKB-PK-130)	−15 mA	18.6 ± 0.7%	This study

Supplementary Materials: The following are available online at http://www.mdpi.com/1996-1073/11/10/2770/s1. Figure S1. Schematic diagram of the electrodialysis reactor used in this study and a photograph. Figure S2. Comparison of propionate transfer through an anion exchange membrane. (A) Amount of propionate transferred to the anodic chamber, (B) applied current in the reactor for 16 h, (C) GC analysis results of the fermentation broth, (D) GC analysis results of the synthetic medium. Figure S3. Membrane and cathodic electrode surface after the completion of electrodialysis for acetate separation. membrane (A) and cathodic electrode (D) from the cell using synthetic broth meida, respectively. (B,E) from centrifuged fermentation broth, (C,F) from non-centrifuged fermentation broth. Figure S4. Estimated current efficiency on different parameters tested. (A) Current efficiency of different currents from −5 mA to −20 mA, (B) Effects of the initial acetate concentration, (C) Effects of different initial anodic pH, (D) Effects of different catholytes with synthetic media (a), centrifuged fermentation broth (b) and non-centrifuged fermentation broth (c).

Author Contributions: Investigation and writing original draft, J.B., C.K.; Investigation, Y.E.S.; Methodology, H.S.I. and M.S.; Supervision, J.R.K.

Funding: This study was supported by the C1 Gas Refinery Program (NRF-2018M3D3A1A01055756) through the National Research Foundation of Korea (NRF) funded by the Ministry of Science, ICT & Future Planning, Korea.

Conflicts of Interest: The authors declare no conflict of interest.

References

1. Ali, J.; Sohail, A.; Wang, L.; Haider, M.R.; Mulk, S.; Pan, G. Electro-microbiology as a promising approach towards renewable energy and environmental sustainability. *Energies* **2018**, *11*, 1822. [CrossRef]
2. Rabaey, K.; Rozendal, R.A. Microbial electrosynthesis—Revisiting the electrical route for microbial production. *Nat. Rev. Microb.* **2010**, *8*, 706. [CrossRef] [PubMed]
3. Im, C.H.; Kim, C.; Song, Y.E.; Oh, S.-E.; Jeon, B.-H.; Kim, J.R. Electrochemically enhanced microbial CO conversion to volatile fatty acids using neutral red as an electron mediator. *Chemosphere* **2018**, *191*, 166–173. [CrossRef] [PubMed]
4. Im, C.H.; Song, Y.E.; Jeon, B.-H.; Kim, J.R. Biologically activated graphite fiber electrode for autotrophic acetate production from CO_2 in a bioelectrochemical system. *Carbon Lett.* **2016**, *20*, 76–80. [CrossRef]
5. Gildemyn, S.; Verbeeck, K.; Slabbinck, R.; Andersen, S.J.; Prévoteau, A.; Rabaey, K. Integrated production, extraction, and concentration of acetic acid from CO_2 through microbial electrosynthesis. *Environ. Sci. Technol. Lett.* **2015**, *2*, 325–328. [CrossRef]
6. Batlle-Vilanova, P.; Puig, S.; Gonzalez-Olmos, R.; Balaguer, M.D.; Colprim, J. Continuous acetate production through microbial electrosynthesis from CO_2 with microbial mixed culture. *J. Chem. Technol. Biotechnol.* **2016**, *91*, 921. [CrossRef]
7. Patil, S.A.; Arends, J.; Vanwonterghem, I.; Van Meerbergen, J.; Guo, K.; Tyson, G.; Rabaey, K. Selective enrichment establishes a stable performing community for microbial electrosynthesis of acetate from CO_2. *Environ. Sci. Technol.* **2015**, *49*, 8833–8843. [CrossRef] [PubMed]
8. LaBelle, E.V.; Marshall, C.W.; Gilbert, J.A.; May, H.D. Influence of acidic pH on hydrogen and acetate production by an electrosynthetic microbiome. *PLoS ONE* **2014**, *9*, e109935. [CrossRef] [PubMed]
9. Srikanth, S.; Singh, D.; Vanbroekhoven, K.; Pant, D.; Kumar, M.; Puri, S.K.; Ramakumar, S.S.V. Electro-biocatalytic conversion of carbon dioxide to alcohols using gas diffusion electrode. *Bioresour. Technol.* **2018**, *265*, 45–51. [CrossRef] [PubMed]
10. del Pilar Anzola Rojas, M.; Zaiat, M.; Gonzalez, E.R.; De Wever, H.; Pant, D. Effect of the electric supply interruption on a microbial electrosynthesis system converting inorganic carbon into acetate. *Bioresour. Technol.* **2018**, *266*, 203–210. [CrossRef] [PubMed]
11. Huang, C.; Xu, T.; Zhang, Y.; Xue, Y.; Chen, G. Application of electrodialysis to the production of organic acids: State-of-the-art and recent developments. *J. Membr. Sci.* **2007**, *288*, 1–12. [CrossRef]
12. Jiang, C.; Chen, H.; Zhang, Y.; Feng, H.; Shehzad, M.A.; Wang, Y.; Xu, T. Complexation Electrodialysis as a general method to simultaneously treat wastewaters with metal and organic matter. *Chem. Eng. J.* **2018**, *348*, 952–959. [CrossRef]
13. Hongo, M.; Nomura, Y.; Iwahara, M. Novel method of lactic acid production by electrodialysis fermentation. *Appl. Environ. Microb.* **1986**, *52*, 314–319.
14. Xue, S.; Wu, C.; Wu, Y.; Chen, J.; Li, Z. Bipolar membrane electrodialysis for treatment of sodium acetate waste residue. *Sep. Purif. Technol.* **2015**, *154*, 193–203. [CrossRef]

15. Patil, R.; Truong, C.; Genco, J.; Pendse, H.; Van Heiningen, A. Applicability of electrodialysis to the separation of sodium acetate from synthetic alkaline hardwood extract. *Tappi J.* **2015**, *14*, 695–708.
16. Kariduraganavar, M.; Nagarale, R.; Kittur, A.; Kulkarni, S. Ion-exchange membranes: Preparative methods for electrodialysis and fuel cell applications. *Desalination* **2006**, *197*, 225–246. [CrossRef]
17. Fidaleo, M.; Moresi, M. Electrodialysis applications in the food industry. *Adv. Food Nutr. Res.* **2006**, *51*, 265–360. [PubMed]
18. Wan, D.; Liu, H.; Qu, J.; Lei, P. Bio-electrochemical denitrification by a novel proton-exchange membrane electrodialysis system—A batch mode study. *J. Chem. Technol. Biotechnol.* **2010**, *85*, 1540–1546. [CrossRef]
19. Mohan, S.V.; Srikanth, S. Enhanced wastewater treatment efficiency through microbially catalyzed oxidation and reduction: Synergistic effect of biocathode microenvironment. *Bioresour. Technol.* **2011**, *102*, 10210–10220. [CrossRef] [PubMed]
20. Chen, X.; Liang, P.; Zhang, X.; Huang, X. Bioelectrochemical systems-driven directional ion transport enables low-energy water desalination, pollutant removal, and resource recovery. *Bioresour. Technol.* **2016**, *215*, 274–284. [CrossRef] [PubMed]
21. Andersen, S.J.; Hennebel, T.; Gildemyn, S.; Coma, M.; Desloover, J.; Berton, J.; Tsukamoto, J.; Stevens, C.; Rabaey, K. Electrolytic membrane extraction enables production of fine chemicals from biorefinery sidestreams. *Environ. Sci. Technol.* **2014**, *48*, 7135–7142. [CrossRef] [PubMed]
22. Długołęcki, P.; Anet, B.; Metz, S.J.; Nijmeijer, K.; Wessling, M. Transport limitations in ion exchange membranes at low salt concentrations. *J. Membr. Sci.* **2010**, *346*, 163–171. [CrossRef]
23. Ghasemi, M.; Daud, W.R.W.; Ismail, M.; Rahimnejad, M.; Ismail, A.F.; Leong, J.X.; Miskan, M.; Liew, K.B. Effect of pre-treatment and biofouling of proton exchange membrane on microbial fuel cell performance. *Int. J. Hydrogen Energy* **2013**, *38*, 5480–5484. [CrossRef]
24. Choi, M.-J.; Chae, K.-J.; Ajayi, F.F.; Kim, K.-Y.; Yu, H.-W.; Kim, C.-W.; Kim, I.S. Effects of biofouling on ion transport through cation exchange membranes and microbial fuel cell performance. *Bioresour. Technol.* **2011**, *102*, 298–303. [CrossRef] [PubMed]
25. Alam, J.; Dass, L.A.; Alhoshan, M.S.; Ghasemi, M.; Mohammad, A.W. Development of polyaniline-modified polysulfone nanocomposite membrane. *Appl. Water Sci.* **2012**, *2*, 37–46. [CrossRef]
26. Upadhyayula, V.K.; Gadhamshetty, V. Appreciating the role of carbon nanotube composites in preventing biofouling and promoting biofilms on material surfaces in environmental engineering: A review. *Biotechnol. Adv.* **2010**, *28*, 802–816. [CrossRef] [PubMed]
27. Kim, C.; Ainala, S.K.; Oh, Y.-K.; Jeon, B.-H.; Park, S.; Kim, J.R. Metabolic flux change in Klebsiella pneumoniae L17 by anaerobic respiration in microbial fuel cell. *Biotechnol. Bioprocess Eng.* **2016**, *21*, 250–260. [CrossRef]
28. Jaime-Ferrer, J.S.; Couallier, E.; Viers, P.; Durand, G.; Rakib, M. Three-compartment bipolar membrane electrodialysis for splitting of sodium formate into formic acid and sodium hydroxide: Role of diffusion of molecular acid. *J. Membr. Sci.* **2008**, *325*, 528–536. [CrossRef]
29. Chukwu, U.; Cheryan, M. Electrodialysis of Acetate Fermentation Broths. In *Twentieth Symposium on Biotechnology for Fuels and Chemicals*; Humana Press: Totowa, NJ, USA, 1999; pp. 485–499.
30. Wei, P.; Xia, A.; Liao, Q.; Sun, C.; Huang, Y.; Fu, Q.; Zhu, X.; Lin, R. Enhancing fermentative hydrogen production with the removal of volatile fatty acids by electrodialysis. *Bioresour. Technol.* **2018**, *263*, 437–443. [CrossRef] [PubMed]
31. Zhang, Y.; Pinoy, L.; Meesschaert, B.; Van der Bruggen, B. Separation of small organic ions from salts by ion-exchange membrane in electrodialysis. *AIChE J.* **2011**, *57*, 2070–2078. [CrossRef]
32. Kim, J.R.; Cheng, S.; Oh, S.-E.; Logan, B.E. Power generation using different cation, anion, and ultrafiltration membranes in microbial fuel cells. *Environ. Sci. Technol.* **2007**, *41*, 1004–1009. [CrossRef] [PubMed]
33. Drews, A. Membrane fouling in membrane bioreactors—Characterisation, contradictions, cause and cures. *J. Membr. Sci.* **2010**, *363*, 1–28. [CrossRef]
34. Zhao, Z.; Shi, S.; Cao, H.; Li, Y.; Van der Bruggen, B. Layer-by-layer assembly of anion exchange membrane by electrodeposition of polyelectrolytes for improved antifouling performance. *J. Membr. Sci.* **2018**, *558*, 1–8. [CrossRef]
35. Lee, H.-M.; Jeon, B.-Y.; Oh, M.-K. Microbial production of ethanol from acetate by engineered Ralstonia eutropha. *Biotechnol. Bioprocess Eng.* **2016**, *21*, 402–407. [CrossRef]
36. López-Garzón, C.S.; Straathof, A.J. Recovery of carboxylic acids produced by fermentation. *Biotechnol. Adv.* **2014**, *32*, 873–904. [CrossRef] [PubMed]

37. Song, Y.E.; Boghani, H.C.; Kim, H.S.; Kim, B.G.; Lee, T.; Jeon, B.-H.; Premier, G.C.; Kim, J.R. Electricity Production by the Application of a Low Voltage DC-DC Boost Converter to a Continuously Operating Flat-Plate Microbial Fuel Cell. *Energies* **2017**, *10*, 596. [CrossRef]
38. Song, Y.E.; Boghani, H.C.; Kim, H.S.; Kim, B.G.; Lee, T.; Jeon, B.H.; Premier, G.C.; Kim, J.R. Maximum Power Point Tracking to Increase the Power Production and Treatment Efficiency of a Continuously Operated Flat-Plate Microbial Fuel Cell. *Energy Technol.* **2016**, *4*, 1427–1434. [CrossRef]
39. Matemadombo, F.; Puig, S.; Ganigué, R.; Ramírez-García, R.; Batlle-Vilanova, P.; Dolors Balaguer, M.; Colprim, J. Modelling the simultaneous production and separation of acetic acid from CO_2 using an anion exchange membrane microbial electrosynthesis system. *J. Chem. Technol. Biotechnol.* **2017**, *92*, 1211–1217. [CrossRef]
40. Bajracharya, S.; Van den Burg, B.; Vanbroekhoven, K.; De Wever, H.; Buisman, C.J.; Pant, D.; Strik, D.P. In situ acetate separation in microbial electrosynthesis from CO_2 using ion-exchange resin. *Electrochim. Acta* **2017**, *237*, 267–275. [CrossRef]
41. Prochaska, K.; Antczak, J.; Regel-Rosocka, M.; Szczygiełda, M. Removal of succinic acid from fermentation broth by multistage process (membrane separation and reactive extraction). *Sep. Purif. Technol.* **2018**, *192*, 360–368. [CrossRef]
42. Wang, X.; Wang, Y.; Zhang, X.; Feng, H.; Xu, T. In-situ combination of fermentation and electrodialysis with bipolar membranes for the production of lactic acid: Continuous operation. *Bioresour. Technol.* **2013**, *147*, 442–448. [CrossRef] [PubMed]

© 2018 by the authors. Licensee MDPI, Basel, Switzerland. This article is an open access article distributed under the terms and conditions of the Creative Commons Attribution (CC BY) license (http://creativecommons.org/licenses/by/4.0/).

Article

Microbial Fuel Cell with Ni–Co Cathode Powered with Yeast Wastewater

Paweł P. Włodarczyk * and Barbara Włodarczyk

Faculty of Natural Sciences and Technology, Institute of Technical Science, University of Opole, Dmowskiego str. 7-9, 45-365 Opole, Poland; barbara.wlodarczyk@uni.opole.pl
* Correspondence: pawel.wlodarczyk@uni.opole.pl; Tel.: +48-077-401-6717

Received: 30 September 2018; Accepted: 14 November 2018; Published: 17 November 2018

Abstract: Wastewater originating from the yeast industry is characterized by high concentration of pollutants that need to be reduced before the sludge can be applied, for instance, for fertilization of croplands. As a result of the special requirements associated with the characteristics of this production, huge amounts of wastewater are generated. A microbial fuel cell (MFC) forms a device that can apply wastewater as a fuel. MFC is capable of performing two functions at the same time: wastewater treatment and electricity production. The function of MFC is the production of electricity during bacterial digestion (wastewater treatment). This paper analyzes the possibility of applying yeast wastewater to play the function of a MFC (with Ni–Co cathode). The study was conducted on industrial wastewater from a sewage treatment plant in a factory that processes yeast sewage. The Ni–Co alloy was prepared by application of electrochemical method on a mesh electrode. The results demonstrated that the use of MFC coupled with a Ni–Co cathode led to a reduction in chemical oxygen demand (COD) by 90% during a period that was similar to the time taken for reduction in COD in a reactor with aeration. The power obtained in the MFC was 6.1 mW, whereas the volume of energy obtained during the operation of the cell (20 days) was 1.27 Wh. Although these values are small, the study found that this process can offer an additional level of wastewater treatment as a huge amount of sewage is generated in the process. This would provide an initial reduction in COD (and save the energy needed to aerate wastewater) as well as offer the means to generate electricity.

Keywords: microbial fuel cell; yeast wastewater; environmental engineering; renewable energy source; cathode; Ni–Co alloy

---

1. Introduction

The food industry plays an important role in the economy of a country [1,2]. It often forms the driving force for other industry branches as well. Due to the fact that huge amounts of wastewater (WW) are generated as a by-product of yeast production, considerable investment is needed for its neutralization [3]. Therefore, attempts are constantly under way to seek the possibility of applying wastewater sludge as a source of energy or as a raw material for further use [4–6]. Studies conducted to this date have demonstrated that it is theoretically possible to derive 9 times more energy from wastewater compared to the energy needed to clean it [7,8]. The yeast industry is one of the branches of the food industry where large amounts of wastewater are generated [9,10]. The yeast industry plays a role not only in the production of baker's yeast but also as a source of substances with various applications in medicine, supplements used to boost the immune system, and in beauty products [11].

As wastewater is formed by a biological mixture, it is characterized by a high content of nitrogen, potassium, and organic substances; low content of heavy metals; and a constant composition (yet relative to the production technology) [12]. For this reason, sewage from the yeast industry can be, and often is, used in agricultural fields (AF) [12,13]. It would be reasonable to apply their enormous potential in the generation of useful energy before redirecting them in the fields. However, for this, the high concentration of COD in yeast wastewater needs to be first reduced. Microbial fuel cells (MFC) form bioelectrochemical systems that can simultaneously pretreat sewage (leading to a decrease in COD) and generate electricity [4,6,14,15]. Therefore, the use of MFCs forms a promising solution that can reduce the concentration of COD coupled with energy production from wastewater [16,17]. MFCs can be powered by both domestic wastewater and processed wastewater [18].

The bacteria that are capable of producing electrons during wastewater treatment play a key role in MFCs [19,20]. Examples of such bacteria include *Geobacter*, *Shewanella*, or *Pseudomonas* genera among others [21–28]. An analysis of reports in this field shows that the highest values of power are generated by MFCs comprising multispecies aggregates, where microorganisms grow in the form of biofilms. Mixed cultures seem to be solid and more efficient compared to single-strain cultures, and their isolation from natural sources is a much easier task. In contrast, the use of pure cultures has technical limitations, mainly due to the need to ensure sterile conditions for growth, and the process usually involves high costs [29].

A microbial fuel cell usually consists of an anode and a cathode separated by a semipermeable proton exchange membrane (PEM), which ensures anaerobic conditions in the anodic area [30]. The activity of an electrode (anode) determines the metabolism of bacteria. In conditions where the anode has a higher potential than other electron mediators available for bacteria, e.g., in the case of sulfides in the source of organic matter, it is the preferred electron mediator [31].

Research on the efficiency of MFCs technology is continuing, and efforts are being made to improve it. An unquestionable goal of research in this field is associated with the development of cells that are capable of producing the greatest volume of electricity with the lowest possible energy delivered into the system [32]. To achieve this, it is important to note the research aimed at increasing the activity of bacteria and electrode materials, e.g., with the use of cathode catalysts. MFCs can be applied to treat domestic wastewater (DWW) as well as industrial wastewater (IWW) [33,34]. The bacteria act as a biocatalyst in microbiological fuel cells [6]; therefore, coal (carbon cloth, carbon brush) is most often used as the anode material [6,35], while carbon materials and metal electrodes perform the role of cathodes. In addition, new electrode materials or catalysts are currently being sought to act as cathodes [36–39].

This paper reports the results of a study into the potential application of wastewater from the yeast industry in a MFC comprising Ni–Co cathode.

2. Materials and Methods

The measurements were carried out by application of wastewater samples from Lesaffre Polska yeast factory. The samples were derived from a wastewater treatment plant (WWTP) located in the factory. The plant policy provides that process wastewater is applied for irrigating farmland, which forms an integral part of the production. However, some of the wastewater from the process goes to a factory wastewater treatment plant with the purpose of ensuring an adequate development of microorganisms needed for the course of water treatment. The volume of process wastewater (PWW) applied at the input of the treatment process depends on many factors, e.g., the total amount of wastewater that enters the treatment plant or the volume of yeast production. Therefore, the waste applied in the treatment plant, which usually contains domestic wastewater and water derived from cleaning equipment, comprises a certain volume of PWW. Table 1 contains a summary of the parameters of WW applied in the measurements.

Table 1. Parameters of wastewater applied for measurements.

Parameter	Value
COD [mg·L^{-1}]	3266
pH	6.5 ± 0.1

The experimental part of the study involved the development of an electrode (cathode), preparation of an experimental setup, measurement of variations in COD concentration, and determination of microbial curves for the fuel cell.

A cathode with a Ni–Co catalyst was applied in the measurements of MFC. The cathode was prepared as a copper mesh with a catalyst coating. The catalyst (Ni–Co alloy) was applied by the electrochemical deposition method [40]. The composition of the mixture and parameters for electrochemical catalyst deposition is summarized in Table 2.

Table 2. Composition of the mixture applied for catalyst deposition (Ni–Co alloy) on the copper mesh.

Component	Volume [g·L^{-1}]	pH	Temperature [K]	Current Density [A·dm^{-2}]	Co Concentration [%]
NiSO$_4$·7H$_2$O	260				
CoSO$_4$·7H$_2$O	14	3.0	293	1.6	15
H$_3$BO$_3$	15				
NiSO$_4$·7H$_2$O	200				
CoSO$_4$·7H$_2$O	20				
H$_3$BO$_3$	30	5.5	293	1.0	25
NaCl	15				
NiSO$_4$·7H$_2$O	195	2.0	293	3.0	50
CoSO$_4$·7H$_2$O	35				
NiSO$_4$·7H$_2$O	140				
CoSO$_4$·7H$_2$O	120				
H$_3$BO$_3$	30	4.0	323	1.0	75
NaCl	15				

The composition of the mixture was selected on the basis of experiments. During the application of the alloy, various concentrations of components (Ni, Co) were obtained (Table 2). The process of preparing the electrode for deposition involved degreasing the copper mesh in KOH solution and its etching in acetic acid and then washing with alcohol [41].

Initially, the study involved the measurements of COD reduction in the examined wastewater. This parameter provides a tool to evaluate the performance of MFCs. The measurements were carried out for three different reactors: one without aeration (Reactor 1), one with aeration (Reactor 2), and one conducted continuously in a working MFC (Reactor 3). An assumption was made that the measurements will be carried out to achieve a 90% decrease in COD concentration [42,43]. In the reactor without aeration (Reactor 1), the wastewater had contact with air only through the wastewater–air interface. In the reactor with aeration (Reactor 2), wastewater was aerated with an air pump with a capacity of 270 L·h^{-1}. In the third reactor (Reactor 3), the wastewater was treated (resulting from the reduction in COD concentration) as a consequence of applying MFC. COD measurements were conducted using the Hanna HI-83224 colorimeter (Hanna Instruments, Woonsocket, RI, USA). In the MFC, carbon cloth was used as the anode material, whereas the cathode was made of a metal mesh with a Ni–Co catalyst. The surface area of the anode was 30 cm^2, while it was 15 cm^2 for the cathode. The cathode was placed in a casing that was printed using 3D technology. The thickness of the single print layers was 0.09 mm. This technology applies high-impact polystyrene (HIPS) material, which is an easy-mold polystyrene that can be applied for producing large items. A Zortrax M200 printer (Zortrax S.A, Olsztyn, Poland) with Z-Suite software (Zortrax S.A, Olsztyn, Poland) was utilized to print the case. One of the walls was removed from the printed housing so as to install a proton exchange membrane (PEM). Nafion PF 117 (The Chemours Company, Wilmington, DE, USA), 183µm thick, was used as the PEM. After installing the PEM, the casing was filled with a catholyte (aqueous KOH solution: 0.1n). Subsequently, a cathode with a Ni–Co catalyst was immersed in the catholyte. The bottom of the casing had perforations and was connected to an air pump (Figure 1). During the operation of the MFC (Reactor 3), the cathode was aerated with the capacity of 10 L·h^{-1}. Figure 1 shows a diagram and image of the casing made of HIPS with the cathode.

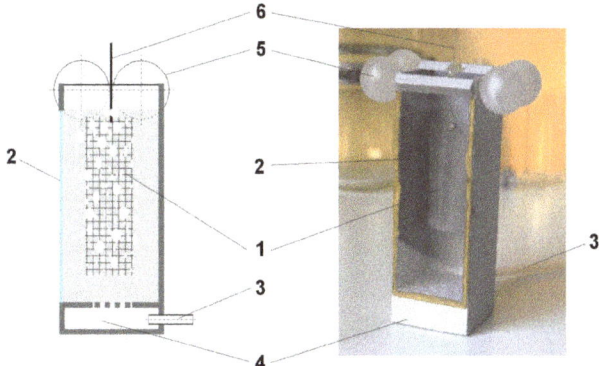

Figure 1. Cathode with casing made of high-impact polystyrene (HIPS): (**a**) cross section (schematic) 1: cathode with Ni–Co catalyst, 2: proton exchange membrane (PEM), 3: air supply, 4: air chamber, 5: float, 6: electrical connection; (**b**) housing with cathode and PEM.

The electrical circuit of the MFC was constantly connected with a 10 Ω resistor. During the work of the MFC (Reactor 3), wastewater was not exchanged or refilled. A constant temperature of 298 K was maintained throughout the measurements. All reactors had the same dimensions (length/width/height: 40cm × 20cm × 20cm). Each reactor (Reactor 1, Reactor 2, and Reactor 3) was filled with wastewater with the volume of 15 L. The third reactor performed the role of MFC because electrodes were installed in it. The measurements of the variations in COD were carried out at one-day intervals. The AMEL System 500 potentiostat (Amel S.l.r., Milano, Italy) with CorrWare software (Scribner Associates Inc, Southern Pines, NC, USA) and the Fluke 8840A multimeter (Fluke Corporation, Everett, WA, USA) was applied for the electrical measurements.

Figure 2 contains a diagram with the design of the third reactor, which played the role of MFC.

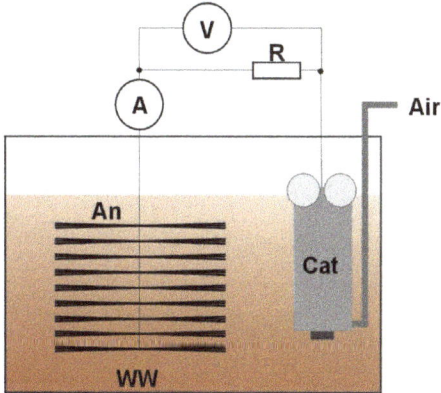

Figure 2. Diagram of microbial fuel cell (MFC) applied in the measurements. A: ammeter, V: voltmeter, R: load, An: anode, Cat: Ni–Co cathode in HIPS housing, WW: wastewater from a wastewater treatment plant in a yeast factory, Air: air supply.

3. Results

All objectives projected for the experimental research were accomplished, i.e., an Ni–Co catalyst was obtained, a reduction in COD for MFCs with cathodes with various percentages of nickel and cobalt was compared, alloy was identified for further tests, a comparison of the decrease in COD was made in the three reactors (one without aeration, one with aeration, and one with MFC), and MFC power curves were determined for all catalysts.

The data summarized in Figure 3 illustrates the reduction in COD concentration accompanying the conditions when MFC was in operation. These measurements were carried out for four catalysts that were obtained, which differed in terms of the nickel and cobalt content (15, 25, 50, and 75% of Co). Wastewater was fed into MFC from the sewage treatment plant located at the yeast factory.

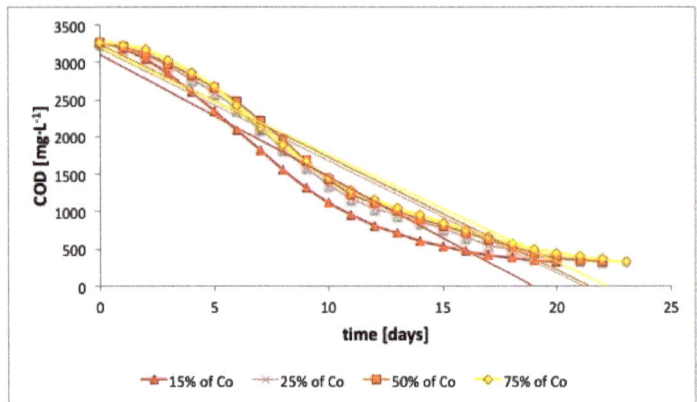

Figure 3. Reduction in chemical oxygen demand (COD) concentration resulting from the operation of our cathode designs with various compositions (15, 25, 50, and 75% Co). Colored curves mark the trend lines.

The trend lines provided grounds for the direct assessment of alloys in terms of the efficiency of COD reduction. On the basis of these results, an alloy with 15% Co content was selected for further measurements.

The curve in Figure 4 shows how the concentration of COD decreased in the three reactors: one without aeration, one with aeration, and one with the working MFC with a Ni–Co cathode (15% Co). All reactors were immersed in wastewater from the wastewater treatment plant located on the factory premises.

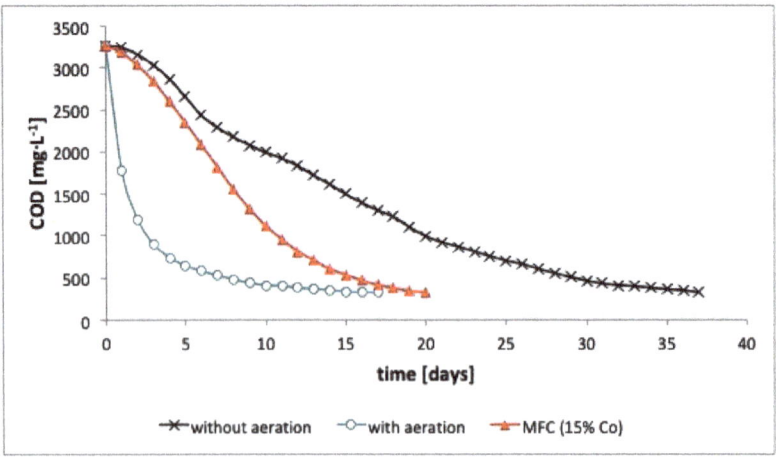

Figure 4. Comparison of the decrease in COD in a reactor with aeration, one without aeration, and one with a working MFC with a Ni–Co cathode (15% Co).

A removal level of COD by 90% was recorded in all reactors (Reactor 1, Reactor 2, and Reactor 3) (Figure 4).

On the basis of the voltage and current measurements in a working MFC, MFC power curves were determined. The MFC power curves derived from wastewater treatment plants in the yeast factory are presented in Figure 5. These curves were determined for four cathode catalysts (Ni–Co), which differed in terms of alloy composition (15, 25, 50, and 75% of Co).

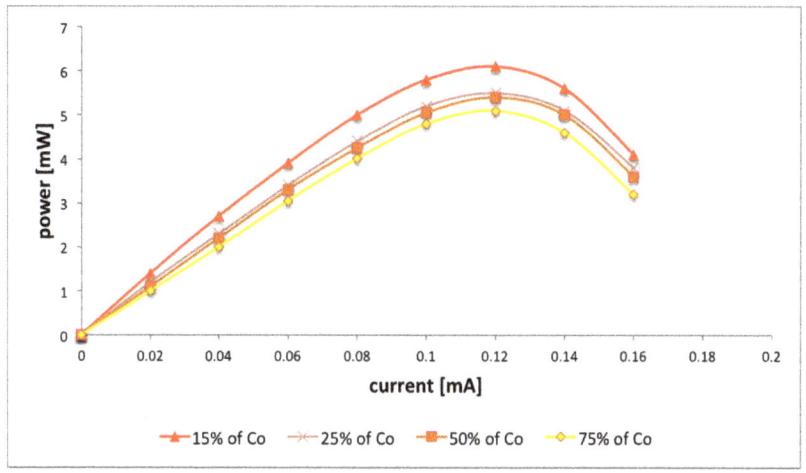

Figure 5. Power curves of microbial fuel cell (MFC) fed by wastewater from a wastewater treatment plant in a yeast factory. The content of Co in the catalyst was equal to 15, 25, 50, and 75%.

4. Discussion

The projected level of the decrease in COD equaled to 90% as recorded in all reactors (Reactor 1, Reactor 2, and Reactor 3). Four Ni–Co alloys were analyzed as the cathode catalysts in the work of MFC (Reactor 3). These alloys differed in terms of the percentage ratio of alloy components (Ni and Co). The characteristics of the curves representing the removal rate of COD for the catalysts comprising 25, 50, and 75% Co were found to be very similar (Figure 3).

However, the curve developed for the drop in COD in the alloy with the 15% Co content took on a different course. This alloy offered the shortest time needed for the reduction in COD, which was equal to 20 days. In addition, the characteristic of the curve for this alloy (marked in red in the figures) was the most beneficial as COD removal was the fastest over the entire range of the measurements. Alloys with 25 and 50% Co ratio provided a 90% reduction in COD over a period of 22 days. However, for the alloy containing 75% Co, the time needed for the adopted COD removal rate was 23 days. Due to the shortest COD removal time (20 days), the alloy comprising 15% Co was selected for further measurements.

The next phase involved a comparison of the efficiency of COD reduction in the following reactor designs: one without aeration (Reactor 1), one with aeration (Reactor 2), and one that played the role of MFC (Reactor 3) (Figure 4). The analysis of these data showed that COD reduction by 90% in the microbiological fuel cell was very similar to the duration of time needed for the COD reduction in the reactor with aeration. Curve characteristics were more favorable for aeration; however, the difference was only three days. It can be noted that for the reduction in COD in the analyzed MFC (Reactor 3), an additional supply of oxygen (from the air) was only necessary for oxygenation of the cathode. This value was 27 times lower than the amount of air supplied to oxygenate Reactor 2.

The power and volume of generated electricity were determined on the basis of these measurements. During the MFC operation, the maximum power was obtained in the range of

5.1–6.1 mW (Figure 5). The maximum volume of energy that was generated by the working MFC comprising a cathode including 15% of Co (20 days) was 1.27 Wh. These values are similar to the results recorded in other MFCs [6,15–18,38,39]. However, when we compare these results with the values recorded in the working MFC (with the Ni–Co cathode) that was fed by municipal sewage, we can note that the use of yeast waste resulted in the production of much larger amount of energy [39,43]. However, it is difficult to compare these results directly as previous studies concerned with Ni–Co catalyst did not have the cathode immersed in the catholyte; moreover, only an alloy comprising 50% Co was analyzed [39,43].

5. Conclusions

The measurements conducted for the purpose of the present study demonstrate that the catalytic activity of the Ni–Co alloy increases along with the increase in the volume ratio of nickel (Ni) in this alloy. Although the power and amount of energy generated by the system are small, the use of MFC can be considered as an additional process of treating sewage, which provides an opportunity to generate small amounts of electricity before the final treatment of wastewater in a treatment plant. It should be noted, however, that the volume of wastewater in a single reactor was equal to only 15 L. In the tested MFC comprising a Ni–Co cathode, the maximum electrical output of 6.1 mW and 1.27 Wh was generated over the period when the cells operated for 20 days. This paper demonstrates the possibility of utilizing Ni–Co alloy (85% Ni, 15% Co) as a cathode catalyst in a working MFC applied for wastewater treatment in a sewage treatment plant in a yeast factory. Therefore, this wastewater can be considered as a potential source of energy to be used for the internal demand of the plant and not only as a problematic type of waste.

Author Contributions: Data curation, B.W.; Investigation, P.P.W. and B.W.; Methodology, P.P.W.; Supervision, P.P.W.

Funding: This research received no external funding.

Acknowledgments: This paper was realized as part of the cooperation between Opole University and Lesaffre Polska yeast factory. This study was possible as part of the scientific internship performed by the authors in Lesaffre Polska factory.

Conflicts of Interest: The authors declare no conflict of interest.

Nomenclature

AF	agricultural fields
An	anode
Cat	cathode
COD	chemical oxygen demand
DWW	domestic wastewater
HIPS	high impact polystyrene
IWW	industrial wastewater
MFC	microbial fuel cell
PEM	proton exchange membrane
PWW	process wastewater
WW	wastewater
WWTP	wastewater treatment plant

References

1. Gallouj, F.; Weber, K.M.; Stare, M.; Rubalcaba, L. The futures of the service economy in Europe: A foresight analysis. *Tech. Forecast. Soc. Chang.* **2015**, *94*, 80–96. [CrossRef]
2. Menrad, K. Market and marketing of functional food in Europe. *J. Food Eng.* **2003**, *56*, 181–188. [CrossRef]
3. Arvanitoyannis, I.S. *Waste Management for the Food Industry*; Elsevier: Amsterdam, The Netherlands, 2010.
4. Cecconet, D.; Molognoni, D.; Callegari, A.; Capodaglio, A.G. Agro-food industry wastewater treatment with microbial fuel cells: Energetic recovery issues. *Int. J. Hydrog. Energy* **2018**, *43*, 500–511. [CrossRef]

5. Barbera, M.; Gurnari, G. *Wastewater Treatment and Reuse in the Food Industry*; Springer: Cham, Switzerland, 2018.
6. Logan, B. *Microbial Fuel Cells*; Wiley: Hoboken, NJ, USA, 2008.
7. Ravindran, R.; Jaiswal, A.K. Exploitation of food industry waste for high-value products. *Trends Biotechnol.* **2016**, *34*, 58–69. [CrossRef] [PubMed]
8. Cusick, R.D.; Kim, Y.; Logan, B.E. Energy capture from thermolytic solutions in microbial reverse-electrodialysis cells. *Science* **2012**, *335*, 1474–1477. [CrossRef] [PubMed]
9. *Utilization of By-Products and Treatment of Waste in the Food Industry*; Oreopoulou, V., Russ, W., Eds.; Springer: New York, NY, USA, 2007.
10. Russ, W.; Meyer-Pittroff, R. Utilizing waste products from the food production and processing industries. *Critical Rev. Food Sci. Nutr.* **2010**, *44*, 57–62. [CrossRef] [PubMed]
11. Muszyńska, B.; Mirosław Malec, M.; Sułkowska-Ziaja, K. Medicinal and cosmetological properties of *Saccharomyces Cerevisiae*. *Postępy Fitoterapii* **2013**, *1*, 54–62.
12. Paruch, A.M.; Pulikowski, K.; Bawiec, A.; Pawęska, K. Assessment of Groundwater Quality in Areas Irrigated with Food Industry Wastewater: A Case of Wastewater Utilisation from Sugar and Yeast Factories. *Envir. Process.* **2017**, *4*, 799–812. [CrossRef]
13. Begea, M.; Berkesy, C.; Berkesy, L.; Cîrîc, A.; Bărbulescu, I.D. Study of the recovery in agriculture of the waste resulted from baker's yeast industry. *Adv. Agric. Botanics* **2017**, *9*, 136–145.
14. Davis, J.B.; Yarbrough, H.F., Jr. Preliminary experiments on a microbial fuel cell. *Science* **1962**, *137*, 615–616. [CrossRef] [PubMed]
15. Franks, A.E.; Nevin, K.P. Microbial fuel cells, a current review. *Energies* **2010**, *3*, 899–919. [CrossRef]
16. Min, B.; Logan, B.E. Continuous electricity generation from domestic wastewater and organic substrates in a at plate microbial fuel cell. *Environ. Sci. Technol.* **2004**, *38*, 5809–5814. [CrossRef] [PubMed]
17. Min, B.; Cheng, S.; Logan, B.E. Electricity generation using membrane and salt bridge microbial fuel cells. *Water Res.* **2005**, *39*, 1675–1686. [CrossRef] [PubMed]
18. Wang, X.; Feng, Y.J.; Lee, H. Electricity production from beer brewery wastewater using single chamber microbial fuel cell. *Water Sci. Technol.* **2008**, *57*, 1117–1121. [CrossRef] [PubMed]
19. Rabaey, K.; Verstraete, W. Microbial fuel cells: Novel biotechnology for energy generation. *Trend. Biotechnol.* **2005**, *23*, 291–298. [CrossRef] [PubMed]
20. Logan, B.E.; Hamelers, B.; Rozendal, R.; Schroder, U.; Keller, J.; Verstraete, W.; Rabaey, K. Microbial Fuel Cells: Methodology and Technology. *Environ. Sci. Technol.* **2006**, *40*, 5181–5192. [CrossRef] [PubMed]
21. Bond, D.R.; Lovley, D.R. Electricity production by *Geobacter sulfurreducens* attached to electrodes. *Appl. Envir. Microbiol.* **2003**, *69*, 1548–1555. [CrossRef]
22. Chaudhuri, S.K.; Lovley, D.R. Electricity generation by direct oxidation of glucose in mediatorless microbial fuel cells. *Nat. Biotechnol.* **2003**, *21*, 1229–1232. [CrossRef] [PubMed]
23. Kim, H.J.; Park, H.S.; Hyun, M.S.; Chang, I.S.; Kim, M.; Kim, B.H. A Mediator-Less Microbial Fuel Cell Using a Metal Reducing Bacterium, *Shewanella Putrefaciens*. *Enzym. Microb. Technol.* **2002**, *30*, 145–152. [CrossRef]
24. Park, H.S.; Kim, B.H.; Kim, H.S.; Kim, H.J.; Kim, G.T.; Kim, M.; Chang, I.S.; Park, Y.K.; Chang, H.I. A novel electrochemically active and Fe(III)-reducing bacterium phylogenetically related to Clostridium butyricum isolated from a microbial fuel cell. *Anaerobe* **2001**, *7*, 297–306. [CrossRef]
25. Pham, C.A.; Jung, S.J.; Phung, N.T.; Lee, J.; Chang, I.S.; Kim, B.H.; Yi, H.; Chun, J. A novel electrochemically active and Fe(III)-reducing bacterium phylogenetically related to Aeromonas hydrophila, isolated from a microbial fuel cell. *FEMS Microbiol. Lett.* **2003**, *223*, 129–134. [CrossRef]
26. Bond, D.R.; Lovley, D.R. Evidence for involvement of an electron shuttle in electricity generation by Geothrix fermentans. *Appl. Environ. Microbiol.* **2005**, *71*, 2186–2189. [CrossRef] [PubMed]
27. Reguera, G.; McCarthy, K.D.; Mehta, T.; Nicoll, J.S.; Tuominen, M.T.; Lovley, D.R. Extracellular electron transfer via microbial nanowires. *Nature* **2005**, *435*, 1098–1101. [CrossRef] [PubMed]
28. Reguera, G.; Nevin, K.P.; Nicoll, J.S.; Covalla, S.F.; Woodard, T.L.; Lovley, D.R. Biofilm and nanowire production leads to incre- ased current in Geobacter sulfurreducens fuel cells. *Appl. Environ. Microbiol.* **2006**, *72*, 7345–7348. [CrossRef] [PubMed]

29. Patil, S.A.; Surakasi, V.P.; Koul, S.; Ijmulwar, S.; Vivek, A.; Shouche, Y.S.; Kapadnis, B.P. Electricity generation using chocolate industry wastewater and its treatment in activated sludge based microbial fuel cell and analysis of developed microbial community in the anode chamber. *Bioresour. Technol.* **2009**, *100*, 5132–5139. [CrossRef] [PubMed]
30. Zhang, X.; Cheng, S.; Huang, X.; Logan, B.E. Improved performance of single-chamber microbial fuel cells through control of membrane deformation. *Biosens. Bioelectron.* **2010**, *25*, 1825–1828. [CrossRef] [PubMed]
31. Aelterman, P.; Freguia, S.; Keller, J.; Verstraete, W.; Rabaey, K. The anode potential regulates bacterial activity in microbial fuel cells. *Appl. Microbiol. Biotechnol.* **2008**, *78*, 409–418. [CrossRef] [PubMed]
32. Juang, D.F.; Lee, C.H.; Hsueh, S.C.; Chou, H.Y. Power generation capabilities of microbial fuel cells with di erent oxygen supplies in the cathodic chamber. *Appl. Biochem. Biotechnol.* **2012**, *167*, 714–731. [CrossRef] [PubMed]
33. Angenent, L.T.; Karima, K.; Al-Dahhan, M.H.; Wrenn, B.A.; Domíguez-Espinosa, R. The wastewater from yeast factory was used in measurements. Production of bioenergy and biochemicals from industrial and agricultural wastewater. *Trend. Biotechnol.* **2004**, *22*, 477–485. [CrossRef] [PubMed]
34. Verstraete, W.; Morgan-Sagastume, F.; Aiyuk, S.; Rabaey, K.; Waweru, M.; Lissens, G. Anaerobic digestion as a core technology in sustainable management of organic matter. *Water Sci. Technol.* **2005**, *52*, 59–66. [CrossRef] [PubMed]
35. Tsai, H.Y.; Wu, C.C.; Lee, C.Y.; Shih, E.P. Microbial fuel cell performance of multiwall carbon nanotubes on carbon cloth as electrodes. *J. Power Source* **2009**, *194*, 199–205. [CrossRef]
36. Dumas, C.; Mollica, A.; Féron, D.; Basséguy, R.; Etcheverry, L.; Bergel, A. Marine microbial fuel cell: Use of stainless steel electrodes as anode and cathode materials. *Electrochim. Acta* **2006**, *53*, 468–473. [CrossRef]
37. Martin, E.; Tartakovsky, B.; Savadogo, O. Cathode materials evaluation in microbial fuel cells: A comparison of carbon, Mn$_2$O$_3$, Fe$_2$O$_3$ and platinum materials. *Electrochim. Acta* **2011**, *58*, 58–66. [CrossRef]
38. Włodarczyk, B.; Włodarczyk, P.P. Microbial fuel cell with Cu-B cathode powering with wastewater from yeast production. *J. Ecol. Eng.* **2017**, *18*, 224–230. [CrossRef]
39. Włodarczyk, B.; Włodarczyk, P.P. Use of Ni-Co alloy as cathode catalyst in single chamber microbial fuel cell. *Ecol. Eng.* **2017**, *18*, 210–216. [CrossRef]
40. Włodarczyk, P.P.; Włodarczyk, B. Possibility of Using Ni-Co Alloy as Catalyst for Oxygen Electrode of Fuel Cell. *Chin. Bus. Rev.* **2015**, *14*, 159–167. [CrossRef]
41. Włodarczyk, P.P.; Włodarczyk, B. Ni-Co alloy as catalyst for fuel electrode of hydrazine fuel cell. *China-USA Bus. Rev.* **2015**, *14*, 269–279. [CrossRef]
42. Huggins, T.; Fallgren, P.H.; Jin, S.; Ren, Z.J. Energy and performance comparison of microbial fuel cell and conventional aeration treating of wastewater. *J. Microb. Biochem. Technol.* **2013**, S6-002. [CrossRef]
43. Włodarczyk, B.; Włodarczyk, P.P.; Kalinichenko, A. Single chamber microbial fuel cell with Ni-Co cathode. In Proceedings of the International Conference Energy, Environment and Material Systems (EEMS 2017), Polanica Zdrój, Poland, 13–15 September 2017.

© 2018 by the authors. Licensee MDPI, Basel, Switzerland. This article is an open access article distributed under the terms and conditions of the Creative Commons Attribution (CC BY) license (http://creativecommons.org/licenses/by/4.0/).

Article

Improved Microbial Electrolysis Cell Hydrogen Production by Hybridization with a TiO₂ Nanotube Array Photoanode

Ki Nam Kim [†], Sung Hyun Lee [†], Hwapyong Kim, Young Ho Park and Su-Il In *

Department of Energy Science and Engineering, DGIST, 333 Techno Jungang-daero, Hyeonpung-myeon, Dalseong-gun, Daegu 42988, Korea; kaizer1354@dgist.ac.kr (K.N.K.); mattlee@dgist.ac.kr (S.H.L.); khp911@dgist.ac.kr (H.K.); nano.e.park@dgist.ac.kr (Y.H.P.)
* Correspondence: insuil@dgist.ac.kr; Tel.: +82-053-785-6417
[†] These authors contributed equally to this work.

Received: 30 August 2018; Accepted: 10 November 2018; Published: 16 November 2018

Abstract: A microbial electrolysis cell (MEC) consumes the chemical energy of organic material producing, in turn, hydrogen. This study presents a new hybrid MEC design with improved performance. An external TiO$_2$ nanotube (TNT) array photoanode, fabricated by anodization of Ti foil, supplies photogenerated electrons to the MEC electrical circuit, significantly improving overall performance. The photogenerated electrons help to reduce electron depletion of the bioanode, and improve the proton reduction reaction at the cathode. Under simulated AM 1.5 illumination (100 mW cm^{-2}) the 28 mL hybrid MEC exhibits a H$_2$ evolution rate of 1434.268 ± 114.174 mmol m^{-3} h^{-1}, a current density of 0.371 ± 0.000 mA cm^{-2} and power density of 1415.311 ± 23.937 mW m^{-2}, that are respectively 30.76%, 34.4%, and 26.0% higher than a MEC under dark condition.

Keywords: microbial electrolysis cell; hydrogen production; TiO$_2$ nanotube

1. Introduction

To curtail global heating, or 'warming' as it is commonly called, and its associated potentially catastrophic climate changes, anthropogenic carbon dioxide emissions should be near zero [1]. Hydrogen is a potential substitute for fossil fuels, and upon combustion produces only water. Hydrogen possesses an energy density of ≈120 MJ kg^{-1}, which compares quite favorably to gasoline at 45.7 MJ kg^{-1}. However, if hydrogen is to be used as a carbon-neutral fuel it must be made on a renewable basis [2], not by the common technique of steam methane reforming (SMR) of natural gas [3]. One approach to carbon-neutral generation of hydrogen is by the use of microbial electrolysis cells (MECs), which convert the chemical energy inherent in wastewater organics into hydrogen. The chemical energy inherent to the organic materials in wastewater is substantial, ≈9.3 times greater than the energy utilized for conventional wastewater treatment [4,5]. MECs offer the dual benefits of both renewable hydrogen generation and wastewater treatment.

A MEC can be viewed as a modified microbial fuel cell (MFC), using the metabolic energy of the inherent organic matter to reduce protons and produce hydrogen [6–8]. MFC exoelectrogens, a type of microorganism, metabolize organic materials of wastewater and generate CO$_2$, H$^+$, and electrons. These electrons can be transferred to the anode in various ways: by an electron mediator [9], direct electron transfer [10], and microbial nanowire [11,12]. The electrons flow through the electrical circuit to reach the air cathode, where they react with atmospheric oxygen and protons to produce water through the oxygen reduction reaction [13,14]. To improve MFC performance, recent studies have investigated the combination of MFCs with external photocurrent sources, such as photosynthetic microorganisms [15,16], a copper oxide photocathode [17], a PtO$_x$-TiO$_2$ composite photocathode [18], a CuInS$_2$ photocathode [19], or a TiO$_2$ nanotube (TNT) array photoanode [20].

In MFC and MEC operation, exoelectrogens attached to the bioanode surface catabolize the organic matter, acetate in this experiment, and generate electrons and protons, and the generated electrons from bioanode flow in the external circuit and reach the cathode [7]. In the MFC, the electrons arrive at the air cathode to react with protons and atmospheric oxygen to produce water, i.e., the oxygen reduction reaction. In a MEC the cathode is not exposed to air, in order to suppress the oxygen reduction reaction, and an external electrical bias is applied to promote proton reduction and thus hydrogen formation [6,21]. The required potential of MEC (>0.2 V) is much lower than the required potential for conventional water electrolysis (1.23 V). Because of this lower required potential, well-controlled MEC can produce hydrogen gas with 2–4 times higher energy than supplied electric energy [22]. There are both single chambered and double chambered MEC, but single chambered MEC without proton exchange membrane has a higher hydrogen production [23]. Recent studies of bioelectrochemical systems have extended to full biological MFC (FB-MFC) [24–26] and full biological MEC (FB-MEC) [27–29], which is composed of biocathode. This replaces expensive chemical catalysts such as Pt, and effectively treats wastewater [30–32]. MEC gets noticeable attention as a sustainable wastewater treatment system with energy recovery [33], and many research teams built pilot scale MEC facilities to validate that MEC can be utilized in real industry [34–39]. In recent studies, various modified MEC structures are suggested for different applications, such as microbial electrodialysis cell (MEDC) [40], microbial reverse-electrodialysis electrolysis cell (MREC) [41,42], microbial electrolysis struvite-precipitation cell (MESC) [43–45], microbial electrolysis desalination and chemical-production cell (MEDCC) [46–49], and microbial saline-wastewater electrolysis cell (MSC) [50–52].

TiO$_2$ is one of the important material for photocatalysis, because of its stability, low cost, and nontoxicity [53]. There is correlation between anatase crystallinity of TiO$_2$ and its photocatalytic activity, especially in nanoparticles [54,55]. In TiO$_2$ photocatalyst, photocatalytic activity is proportionate to the fraction of crystallized anatase [56]. This is because anatase TiO$_2$ has longer carrier lifetimes [57,58] and faster electron transfer rates [59] than amorphous or rutile TiO$_2$. To achieve a maximum photocatalytic efficiency of TiO$_2$, maximizing surface area is important, therefore nanoparticulated TiO$_2$ is widely utilized. 1D nanostructure TiO$_2$ such as nanotube has crucial advantages over other nanostructures, because we can control its physical properties, specifically diameter and length, which results in an improved and effective system [60]. In a previous study on a MFC coupled with an external TNT array photoanode, we confirmed that supplementary electrons accelerate the oxygen reduction reaction at air cathode, improving overall MFC performance [20].

In this study, we couple an external TNT array photoanode, made by electrochemical anodization of Ti foil, with a single chamber MEC. Under simulated 1.0 sun solar light (AM 1.5, 100 mW cm^{-2}) the 28 mL hybrid MEC demonstrates, in comparison with a MEC under dark, a significantly greater rate of H$_2$ evolution rate (1434.268 ± 114.174 mmol m^{-3} h^{-1}), current density (0.371 ± 0.000 mA cm^{-2}), and power density (1415.311 ± 23.937 mW m^{-2}). It appears that the additional photoelectrons from the TNT array photoanode significantly reduce electron depletion of the bioanode, and accelerate the proton reduction reaction at the cathode, and hence overall MEC performance.

2. Materials and Methods

2.1. Chemicals and Materials

The Pt-catalyzed cathode was made using: Carbon cloth (Fuel Cell Earth LLC, Woburn, MA, USA), polytetrafluoroethylene (PTFE, 60 wt.% dispersion in H$_2$O, Sigma-Aldrich, St. Louis, MO, USA), Pt on Vulcan XC-72 (10%, Premetek Co., Wilmington, DE, USA), carbon vulcan powder (Premetek, Wilmington, DE, USA), Nafion perfluorinated resin solution (15–20%, Sigma-Aldrich, St. Louis, MO, USA), and 2-propanol (99.5%, Sigma-Aldrich, St. Louis, MO, USA). A carbon brush (mill-rose, Mentor, OH, USA) was used to make the bioanode. Disodium phosphate (anhydrous, for molecular biology, Applichem, Darmstadt, Germany), potassium chloride (Sigma-Aldrich, St. Louis, MO, USA), sodium phosphate monobasic monohydrate (ACS reagent, ≥98%, Sigma-Aldrich, St. Louis, MO, USA), ammonium chloride (for molecular biology, suitable for cell culture, ≥99.5%, Sigma-Aldrich, St. Louis, MO, USA), and sodium acetate (anhydrous, for molecular biology, ≥99%, Sigma-Aldrich, St. Louis, MO, USA) were used to make the MFC and MEC medium. Titanium foil (0.1 mm thickness, 99.5%, Nilaco Co., Tokyo, Japan), carbon paper (CNL energy, Seoul, Korea), ammonium fluoride (98.0%, Alfa Aesar, Haverhill, MA, USA), and ethylene glycol (Spectrophotometric grade, 99%, Alfa Aesar, Haverhill, MA, USA) were employed to synthesize the TNT array photoanode. Titanium wire (1.0 mm diameter, 99.5%, Nilaco Co., Tokyo, Japan) was used to connect the cathode with the external electric circuit. A 100 W Xenon solar simulator (LCS-100, Oriel® Instruments, Irvine, CA, USA) with an AM 1.5 filter was employed as the light source in an otherwise dark room. External bias application, and electrical and electrochemical characterizations were implemented by use of a potentiostat (Bio-Logic, VSP model, Seyssinet-Pariset, France).

2.2. Synthesis of TNT Array Photoanodes

The TNT array photoanode was synthesized by electrochemical anodization of Ti foil (6.0 cm × 4.0 cm). Prior to anodization the Ti foil sample was washed by sonication in acetone, ethanol, and DI water (deionized water), for 10 min each. Electrochemical anodization was conducted by a two-electrode cell using Ti foil as working electrode and carbon paper as counter electrode and interval between both electrodes was 2 cm. The anodization electrolyte was ethylene glycol with 0.5 wt.% NH_4F and 2.0 vol.% DI water. Anodization was performed at 40 V for 30 min, as previously reported [20,61]. With anodization, Ti^{4+} ions are generated (Ti → Ti^{4+} + $4e^-$) and driven from Ti substrate to the electrolyte by electric field. Also, hydroxyl ions (OH^-) and O^{2-} anions are formed due to water of the anodic oxide film in the electrolyte [62]. These anions form TiO_2 oxide through several reactions (Ti^{4+} + $2O^{2-}$ → TiO_2; Ti^{4+} + $4OH^-$ → $Ti(OH)_4$; $Ti(OH)_4$ → TiO_2 + $2H_2O$). F^- ions of electrolyte can combine with Ti^{4+} ions to generate TiF_6^{2-} (Ti^{4+} + $6F^-$ → TiF_6^{2-}). Because Gibb's free energy of TiF_6^{2-} ($\triangle G°_{298}$ = −2118.4 kJ mol^{-1}) is much lower than Gibb's free energy of TiO_2 ($\triangle G°_{298}$ = −821.3 kJ mol^{-1}) [63], F^- ion can also dissolve the oxide and hydrated layer, which results as etching of the foil (TiO_2 + $6F^-$ + $4H^+$ → TiF_6^{2-} + H_2O; $Ti(OH)_4$ + $6F^-$ → TiF_6^{2-} + $4OH^-$). Through competition between oxidation reactions of Ti and dissolution reactions of Titanium oxide, TiO_2 nanotube array structure is produced [64,65]. The anodized Ti foil was sonicated in ethanol for 2 min to wash contaminants and debris off the anodized Ti foil surface and then annealed at 450 °C for 2 h. Through annealing process, the amorphous anodized TiO_2 nanotube array is converted to crystalline anatase phase [64,66]. The crystallinity of the TNT array was investigated by X-ray diffraction spectroscopy (XRD, Empyrean, Panalytical, Almelo, The Netherland) with Cu Kα radiation (λ = 1.54 Å) as an X-ray source, operating at 40 kV and 30 mA, scanned in the range of 2θ = 20–80° with a rate of 2.5° min^{-1}. Morphological analysis was performed by field emission scanning electron microscope (FE-SEM) (S-4800, Hitachi, Tokyo, Japan). The TNT array sample was cut into circular form with a diameter of 3.0 cm (7.0 cm^2 area) for attaching with the MEC. We measured the stability of TNT array by changes in photocurrent changes with respect to time. The photocurrent was measured by potentiostat (Bio-Logic, VSP model, Seyssinet-Pariset, France), with TNT array serving as the working electrode, the reference electrode was Ag/AgCl reference electrode (3.0 M KCl, EC-frontier, Kyoto, Japan), and Pt wire was used as the counter electrode. Electrolyte was composed of phosphate buffer solution with sodium acetate, which has been used as MFC and MEC medium (Composition is mentioned in Section 2.3). A potential of 0.40 V was applied and the TNT array was illuminated by a 100 W Xenon solar simulator (AM 1.5) for 60 h.

2.3. MFC Operation for Bioanode Preparation

The MEC bioanode was obtained by enriching an electroactive biofilm of exoelectrogen on the surface of a carbon brush anode, using a single chamber MFC (cylindrical reaction chamber of 4 cm length, 7 cm^2 cross section, 28 mL volume) [67], comprised of a Pt-catalyzed cathode, carbon brush anode, and external resister (1000 Ω), see Figure S1. To make Pt-catalyzed cathode, used air cathode of MFC and cathode of MEC, in which one side of carbon cloth, contact with MFC and MEC medium, was coated with a mixture of Pt on Vulcan XC-72 and Nafion resin (0.5 mg cm^{-2} Pt), and the other side of carbon cloth, contact with air in MFC, was coated with a layer of 40% PTFE (polytetrafluoroethylene) with carbon black and four layers of 60% PTFE [68]. To make the ammonia treated carbon brush anode, carbon brush was washed with acetone and DI water, each, to remove contaminants prior to ammonia gas treatment. Later, the carbon brush was treated in 5% NH$_3$ (He base) at 700 °C for 60 min to improve the performance of bioanode [69].

The MFCs were charged with sludge of an anaerobic reactor, source of the exoelectrogen, collected from Daegu Environmental Corporation Seobu sewage treatment plant (Daegu, Korea). The medium contains sodium acetate (2 g L^{-1}) as a substrate and a 50 mM phosphate buffer solution (PBS), the composition of which is 4.58 g L^{-1} Na$_2$HPO$_4$; 0.13 g L^{-1} KCl; 2.45 g L^{-1} NaH$_2$PO$_4$·H$_2$O; 0.31 g L^{-1} NH$_4$Cl; trace vitamins (5 mL L^{-1}) and minerals (12.5 mL L^{-1}) stock solutions [70]. The MFC medium was replaced by fresh medium every two days. The MFC was operated for 10 weeks for the formation of a mature exoelectrogenic biofilm on the carbon brush anode and to show stable performance.

2.4. Hybrid MEC Fabrication and Operation

The schematic design of the hybrid MEC is shown in Figure 1. The following steps were taken to convert the MFC to a MEC: (1) The Pt-catalyzed cathode of the MFC was covered with an acrylic plate to prevent oxygen from reaching the cathode. With the oxygen reduction reaction blocked, hydrogen is evolved from the Pt-catalyzed cathode. (2) The carbon brush bioanode was separated from the reactor and exposed to air for 40 min to suppress methanogen growth [7]. After aerial exposure, the bioanode was re-assembled into the reactor. (3) The reactor chamber was filled with medium, and the whole purged with ultra high purity (UHP) nitrogen for 35 min. The external resistor of the MEC was 10 Ω. (4) The TNT array photoanode was electrically connected, see Figure 1, to the bioanode of the MEC.

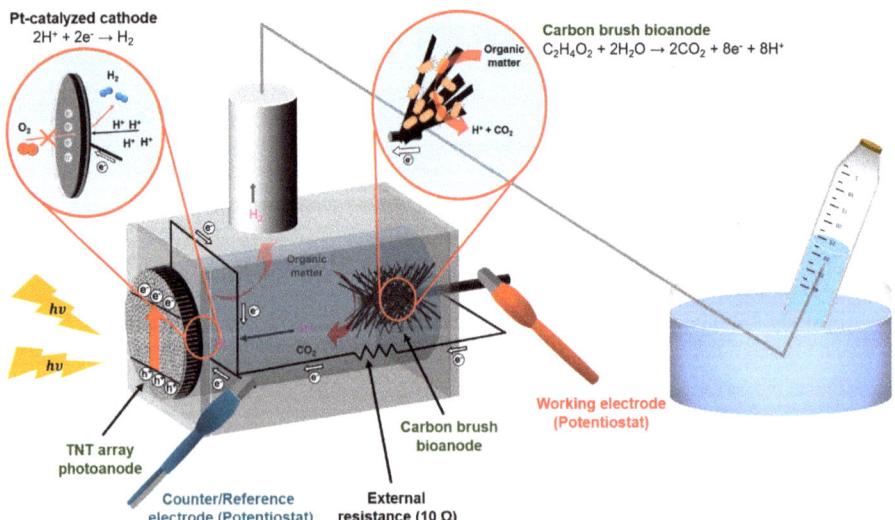

Figure 1. Schematic illustration of the hybrid microbial electrolysis cell (MEC) with external TiO$_2$ photoanode.

The TNT array photoanode was illuminated by a 100 W Xenon solar simulator with an AM 1.5 filter for 24 h. A potentiostat (Bio-Logic, VSP model, Seyssinet-Pariset, France) was used to apply a 0.40 V bias between bioanode and Pt-catalyzed cathode. The electrode potential was measured using a three-electrode system against an Ag/AgCl reference electrode (3.0 M KCl, EC-frontier, Kyoto, Japan). A conical tube with a water-filled rubber septum was connected to the MEC for capturing generated gas which, after 24 h MEC operation, was analyzed using a gas chromatograph (Agilent, 6890N, Santa Clara, CA, USA) using a 19095-MOSE column (30 m × 0.53 mm i.d., 50.0 μm film thickness). GC data to evaluate the hydrogen evolution were taken three times (n = 3) for the hybrid MEC under illumination and the MEC under dark.

2.5. Hybrid MEC Performance Evaluation

The MEC performance was evaluated by polarization curves and power density curves. To measure the current density of MEC, potentiostat (Bio-Logic, VSP model, Seyssinet-Pariset, France) was set in chronopotentiometry mode, and the working electrode was connected with cathode, the reference electrode and the counter electrode were connected with bioanode. The applied current started from −200 μA and was increased by −200 μA until potential went lower than 0 V. There were no external resistance and any external bias; a similar condition for measuring the current density of MFC. MEC current density (mA cm^{-2}) and power density (mW cm^{-2}) were normalized to the cathode area (7 cm^2). Power Density = I · V · A^{-1}, where I = current, V = voltage, and A = cathode area. The MEC polarization curve was measured with the TNT array photoanode illuminated (100 W Xenon solar simulator, AM 1.5 filter). All measurements were taken three times (n = 3) for evaluating the hybrid MEC under illumination and the MEC under dark.

3. Results and Discussion

The TiO$_2$ photoanode, fabricated by oxidation of Ti foil, is a self-organized array of nanotubes resting upon the underlying Ti foil substrate [71,72]. Figure 2a is a FE-SEM image of the top surface, with Figure 2b showing sample cross-section. As seen, the TiO$_2$ photoanode is comprised of well-aligned nanotubes of 4.04~4.35 μm length.

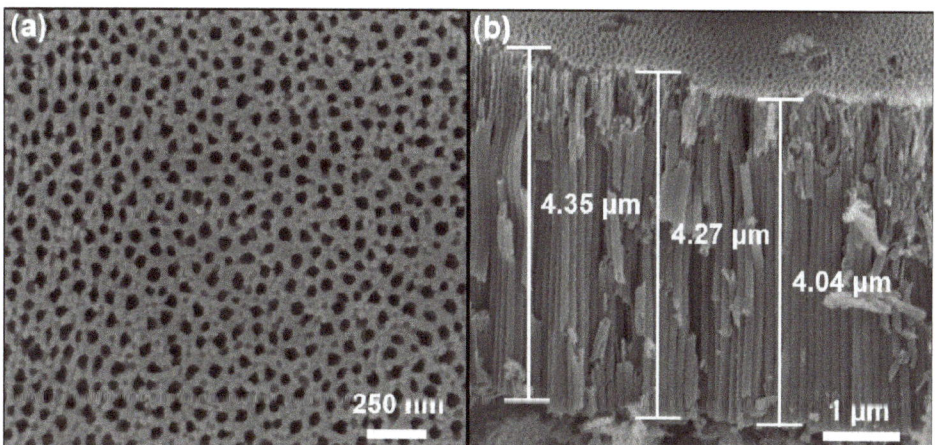

Figure 2. (a) Surface and (b) cross-sectional field emission scanning electron microscope (FE-SEM) images of TiO$_2$ nanotube (TNT).

The crystal structures of Ti foil and TNT arrays revealed by X-ray diffraction (XRD) patterns are shown in Figure 3. The Ti foil showed peaks at 2θ values of 40.16°, 52.98°, 70.63° and 76.19° [73,74]. The unannealed TNT array sample showed peaks similar to those of the Ti foil, but also a slight peak at 2θ = 62.75°, which can be referred to (204) plane of anatase TiO$_2$. The annealed TNT arrays showed 2θ peaks at 25.39°, 36.91°, 37.97°, 48.13°, 54.10°, 55.13°, 62.75°, 68.86° and 75.12°. These peaks are corresponding, respectively, to (101), (103), (004), (200), (105), (211), (204), (116) and (215) planes of anatase [75]. It proves that the annealing process makes amorphous unannealed TNT array as crystalline anatase phase.

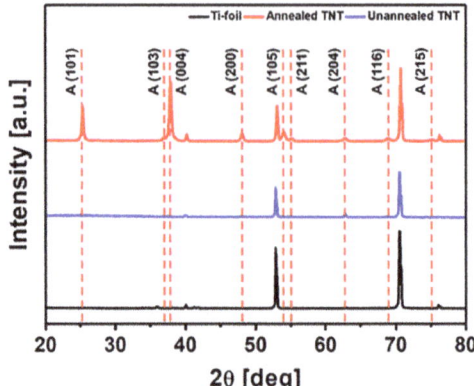

Figure 3. XRD patterns of Ti foil (black), unannealed (blue) and annealed (red) TNT array.

The photocurrent measurement of TNT array's photoanode with respect to time is shown in Figure S2. After illuminating for 2 min, the photocurrent of TNT array photoanode increased rapidly and reached a maximum photocurrent density of 0.11088 mA cm^{-2}. This was followed by a fall in photocurrent density to 0.10212 mA cm^{-2}; a value obtained after 13 min of illumination. Beyond this, a gradual decrease in the photocurrent was observed (-0.00391 mA cm^{-2} h^{-1} to -0.00049 mA cm^{-2} h^{-1}). Photocurrent density reaches 90% (0.10077 mA cm^{-2}), 80% (0.08877 mA cm^{-2}), 75% (0.08322 mA cm^{-2}), and 50% (0.055578 mA cm^{-2}) with respect to maximum photocurrent density in 45 min, 6.2 h, 12.4 h, 52.4 h after illumination, respectively. Photocurrent density after 24 h illumination is 66.3% (0.07351 mA cm^{-2}) of maximum value.

The power density and I–V polarization curves of a hybrid MEC, illuminated and in the dark (equal to a MEC without photoanode) are shown in Figure 4a. It can be observed that under simulated solar light the hybrid MEC showed a current density of 0.371 ± 0.000 mA cm^{-2} ($n = 3$) and power density of 1415.311 ± 23.937 mW m^{-2} ($n = 3$), while in the dark exhibited a current density of 0.276 ± 0.010 mA cm^{-2} ($n = 3$) and power density of 1122.848 ± 32.664 mW m^{-2} ($n = 3$). Coupling of a TNT array photoanode to the MEC resulted in a 34.4% improvement in current, and 26.0% increase in power density.

The open circuit and working electrode potential of the hybrid MEC, under illumination and under dark, are shown in Figure 4b. The cathode polarization curves of both hybrid MECs under illumination/under dark have similar slopes, while for the anode polarization curves there is considerable difference. For the MEC under dark, the slope of the anode potential curve rapidly increases after 0.295 ± 0.019 mA cm^{-2} ($n = 3$), while in the hybrid MEC under illumination the slope of the anode potential curve remains almost constant until 0.533 ± 0.053 mA cm^{-2} ($n = 3$). This sudden change in the anode potential curve of the MEC under dark is similar with "power overshoot" behavior [76,77]. Generally, power overshoot occurs when the anodic overpotential caused by electron depletion overruns the cathodic overpotential [78]. However, in our case, this sudden change arises because of electron depletion at the bioanode as given in Figure 4b. Nonetheless, this lack of electrons at the bioanode is compensated by additional electrons supplied by the TNT array photoanode and thus the overall electron production of the anodes (TNT photoanode + bioanode) exceeds electron reduction at the cathode. This delays the electron depletion at bioanode, and results in increased power density. Figure S3 shows the power density, I–V polarization, and electrode potential curves for hybrid MEC under illumination and MEC under dark in order to check reproducibility.

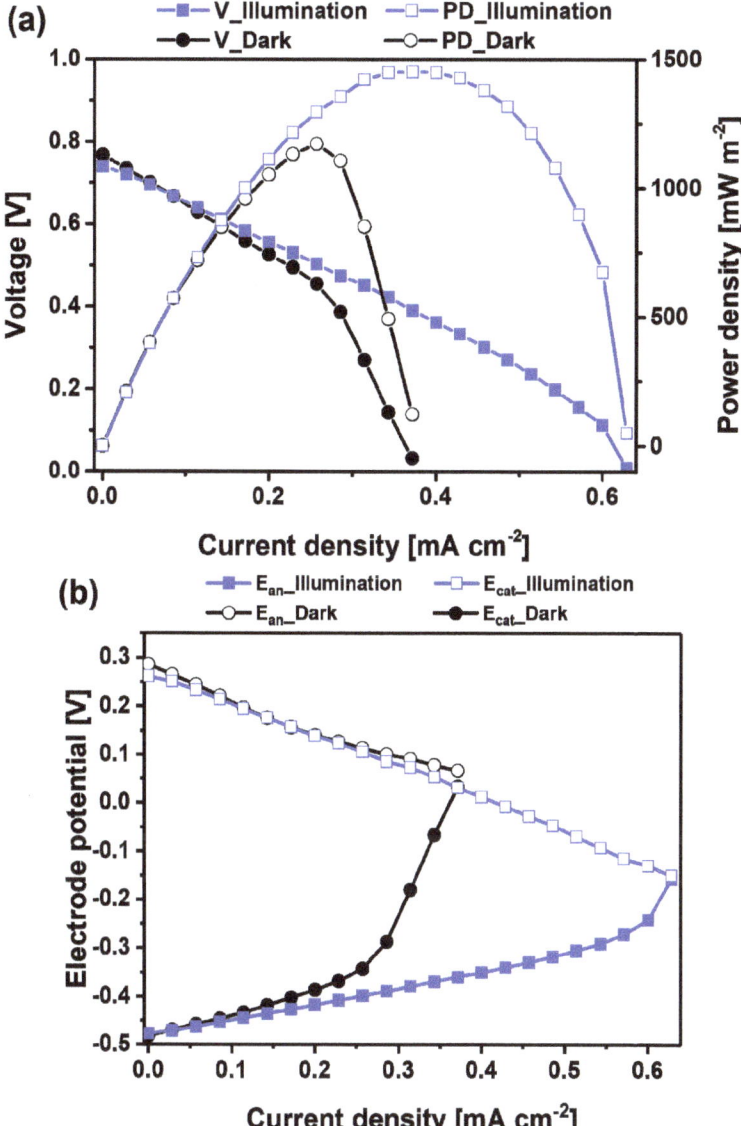

Figure 4. (**a**) Power density (open symbol), I–V polarization (solid symbol), and (**b**) electrode potential of hybrid MEC under simulated solar light and under dark (equal to a MEC without photoanode).

The quantity of hydrogen produced, measured by the GC, in ppm h^{-1} and µmol h^{-1} was calculated as shown:

$$H_2 \: [] = \frac{\mu mol \: of \: H_2 \: produced}{mol \: of \: gaseous \: mixture} \times \frac{1}{t[h]} \quad (1)$$

$$H_2 \: [\mu mol \: h^{-1}] = \frac{\mu mol \: of \: H_2 \: produced}{mol \: of \: gaseous \: mixture} \times \frac{1}{t[h]} \times \frac{1}{V\left[mL \: mol^{-1}\right]} \times Total \: produced \: gas \: [mL] \quad (2)$$

where "V" is the molar volume (22,400 mL mol^{-1}) of gas at STP and "t" is total reaction time.

$$H_2 \left[\text{mmol m}^{-3} \text{ h}^{-1} \right] = \frac{H_2 \left(\mu\text{mol h}^{-1} \right)}{V_R \text{ [mL]}} \times \frac{10^{-3} \text{ mmol}}{1 \text{ }\mu\text{mol}} \times \frac{1 \text{ mL}}{10^{-6} \text{ m}^3} \quad (3)$$

where "V_R" is volume of reactor (28 mL).

Under simulated solar light the hybrid MEC displayed an H_2 evolution rate of 40.159 ± 3.197 µmol h^{-1} (n = 3), while MEC under dark showed 30.711 ± 0.431 µmol h^{-1} (n = 3). By normalizing against reactor volume, H_2 evolution rate is translated as 1434.268 ± 114.174 mmol m^{-3} h^{-1} (n = 3) and 1096.809 ± 15.388 mmol m^{-3} h^{-1} (n = 3), as shown in Figure 5.

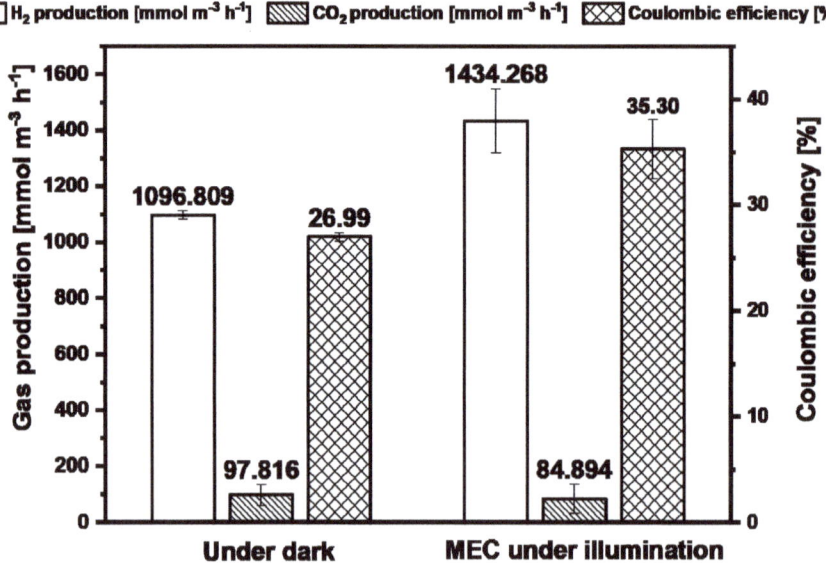

Figure 5. H_2 production rate, CO_2 production rate, and coulombic efficiency by the MEC under dark condition and hybrid MEC under simulated solar light for 24 h.

The coulombic efficiency (CE) of the MEC can be calculated as:

$$CE = \frac{C_P}{C_{Ti}} \times 100\% \quad (4)$$

where "C_P" is the total coulombs calculated by the amount of hydrogen produced, and "C_{Ti}" is the theoretical coulombs that can be produced from wastewater substrate. C_P and C_{Ti} can be calculated via following equations:

$$C_P \text{ (C)} = n \times F \times H_2 \text{ evolution rate} \times Operation \text{ } time \quad (5)$$

$$C_{Ti} \text{ (C)} = \frac{FbSv}{M} \quad (6)$$

where "n" is number of moles of electrons per mole of reduced hydrogen, "F" is Faraday's constant, "b" is number of moles of produced electrons per mole of oxidized substrate, "S" is substrate concentration of medium, "v" is medium volume, and "M" is molecular weight of the substrate [67]. Considering substrate oxidation equation of bioanode ($C_2H_4O_2 + 2H_2O \rightarrow 2CO_2 + 8e^- + 8H^+$) and hydrogen reduction equation of cathode (2H$^+$ + 2e$^-$ → H$_2$), calculations of C_P and C_{Ti} of MEC appear as the following equation:

$$C_{P,illumination} = 2 \ electrons \times 96485 \ \frac{C}{mol} \times 40.159 \ \frac{\mu mol}{h} \times 24 \ h = 185.99 \ C \quad (7)$$

$$C_{P,dark} = 2 \ electrons \times 96485 \ \frac{C}{mol} \times 30.711 \ \frac{\mu mol}{h} \times 24 \ h = 142.23 \ C \quad (8)$$

$$C_{Ti} = \frac{96485 \ \frac{C}{mol} \times 8 \ electrons \times 2 \ \frac{g}{L} \times 0.028 \ L}{82.0343 \ \frac{g}{mol}} = 526.92 \ C \quad (9)$$

Under simulated solar light, the coulombic efficiency of the hybrid MEC is 35.30 ± 2.81% (n = 3), while that of the MEC under dark is 26.99 ± 0.38% (n = 3), shown in Figure 5.

4. Conclusions

Herein we combine a MEC with a TNT array photoanode to significantly improve H$_2$ production. The hybrid MEC design achieves a H$_2$ evolution rate of 1434.268 ± 114.174 mmol m^{-3} h^{-1} (n = 3) and coulombic efficiency of 35.30 ± 2.81% (n = 3) under simulated 1.0 sun solar light (AM 1.5, 100 mW cm^{-2}), values some 30.76% higher than a MEC under dark. The hybrid MEC achieved a current density of 0.371 ± 0.000 mA cm^{-2} (n = 3) and power density of 1415.311 ± 23.937 mW m^{-2} (n = 3) under illumination, values, respectively, 34.4% and 26.0% above those of a MEC under dark. We believe the enhanced performance is due to an increased quantity of electrons arriving at the bioanode that, in turn, accelerates the proton reduction reaction. We suggest this photocatalytic hybridization strategy will improve the efficiency of MEC hydrogen production and it will help to make hydrogen a true carbon-neutral energy source, which can solve anthropogenic carbon dioxide emission problem and global warming.

Supplementary Materials: The following are available online at http://www.mdpi.com/1996-1073/11/11/3184/s1, Figure S1: Schematic illustration of the MFC for preparing matured bioanode, Figure S2: Photocurrent dependent time of the TNT array photoanode under simulated solar light for 60 h, Figure S3: (a) Power density, I–V polarization, and (b) electrode potential curves for hybrid MEC under illumination and MEC under dark for checking reproducibility.

Author Contributions: Conceptualization, S.-I.I.; methodology, S.H.L. and Y.H.P.; validation, K.N.K., H.K. and S.H.L.; formal analysis, H.K. and S.H.L.; investigation, K.N.K., H.K. and S.H.L.; data curation, K.N.K., H.K., S.H.L. and Y.H.P.; writing—original draft preparation, K.N.K.; writing—review and editing, K.N.K., H.K. and S.-I.I.; visualization, K.N.K., S.H.L., H.K. and Y.H.P.; supervision, S.-I.I.; project administration, S.-I.I.; funding acquisition, S.-I.I.

Funding: This research was supported by the Technology Development Program to Solve Climate Changes of the National Research Foundation (NRF) and funded by the Ministry of Science, ICT & Future Planning (2015M1A2A2074670).

Conflicts of Interest: The authors declare no conflicts of interest.

References

1. Matthews, H.D.; Caldeira, K. Stabilizing climate requires near-zero emissions. *Geophys. Res. Lett.* **2008**, 35. [CrossRef]
2. Grimes, C.; Varghese, O.; Ranjan, S. *Light, Water, Hydrogen: The Solar Generation of Hydrogen by Water Photoelectrolysis*; Springer: New York, NY, USA, 2008; ISBN 978-0-387-33198-0.
3. Satyapal, S. Hydrogen and Fuel Cells Overview, DLA Worldwide Energy Conference, National Harbor, MD. 2017. Available online: https://www.energy.gov/eere/fuelcells/downloads/hydrogen-and-fuel-cells-overview (accessed on 14 August 2018).
4. Logan, B.E. *Microbial Fuel Cells*; John Wiley & Sons, Inc.: New York, NY, USA, 2008; pp. 6–7, ISBN 9780470239483.

5. Shizas, I.; Bagley, D.M. Experimental determination of energy content of unknown organics in municipal wastewater streams. *J. Energy Eng.* **2004**, *130*, 45–53. [CrossRef]
6. Liu, H.; Grot, S.; Logan, B.E. Electrochemically assisted microbial production of hydrogen from acetate. *Environ. Sci. Technol.* **2005**, *39*, 4317–4320. [CrossRef] [PubMed]
7. Call, D.; Logan, B.E. Hydrogen production in a single chamber microbial electrolysis cell lacking a membrane. *Environ. Sci. Technol.* **2008**, *42*, 3401–3406. [CrossRef] [PubMed]
8. Rozendal, R.A.; Hamelers, H.V.M.; Euverink, G.J.W.; Metz, S.J.; Buisman, C.J.N. Principle and perspectives of hydrogen production through biocatalyzed electrolysis. *Int. J. Hydrogen Energy* **2006**, *31*, 1632–1640. [CrossRef]
9. Rabaey, K.; Boon, N.; Siciliano, S.D.; Verhaege, M.; Verstraete, W. Biofuel cells select for microbial consortia that self-mediate electron transfer. *Appl. Environ. Microbiol.* **2004**, *70*, 5373–5382. [CrossRef] [PubMed]
10. Bond, D.R.; Lovley, D.R. Electricity production by Geobacter sulfurreducens attached to electrodes. *Appl. Environ. Microbiol.* **2003**, *69*, 1548–1555. [CrossRef] [PubMed]
11. Gorby, Y.A.; Yanina, S.; McLean, J.S.; Rosso, K.M.; Moyles, D.; Dohnalkova, A.; Beveridge, T.J.; Chang, I.S.; Kim, B.H.; Kim, K.S. Electrically conductive bacterial nanowires produced by Shewanella oneidensis strain MR-1 and other microorganisms. *Proc. Natl. Acad. Sci. USA* **2006**, *103*, 11358–11363. [CrossRef] [PubMed]
12. Reguera, G.; McCarthy, K.D.; Mehta, T.; Nicoll, J.S.; Tuominen, M.T.; Lovley, D.R. Extracellular electron transfer via microbial nanowires. *Nature* **2005**, *435*, 1098–1101. [CrossRef] [PubMed]
13. Min, B.; Logan, B.E. Continuous electricity generation from domestic wastewater and organic substrates in a flat plate microbial fuel cell. *Environ. Sci. Technol.* **2004**, *38*, 5809–5814. [CrossRef] [PubMed]
14. Logan, B.E.; Regan, J.M. Electricity-producing bacterial communities in microbial fuel cells. *Trends Microbiol.* **2006**, *14*, 512–518. [CrossRef] [PubMed]
15. Strik, D.P.; Timmers, R.A.; Helder, M.; Steinbusch, K.J.J.; Hamelers, H.V.M.; Buisman, C.J.N. Microbial solar cells: Applying photosynthetic and electrochemically active organisms. *Trends Biotechnol.* **2011**, *29*, 41–49. [CrossRef] [PubMed]
16. Qi, X.; Ren, Y.; Liang, P.; Wang, X. New insights in photosynthetic microbial fuel cell using anoxygenic phototrophic bacteria. *Bioresour. Technol.* **2018**, *258*, 310–317. [CrossRef] [PubMed]
17. Qian, F.; Wang, G.; Li, Y. Solar-driven microbial photoelectrochemical cells with a nanowire photocathode. *Nano Lett.* **2010**, *10*, 4686–4691. [CrossRef] [PubMed]
18. Ansari, S.A.; Khan, M.M.; Ansari, M.O.; Cho, M.H. Improved electrode performance in microbial fuel cells and the enhanced visible light-induced photoelectrochemical behaviour of PtOx@ M-TiO$_2$ nanocomposites. *Ceram. Int.* **2015**, *41*, 9131–9139. [CrossRef]
19. Wang, S.; Yang, X.; Zhu, Y.; Su, Y.; Li, C. Solar-assisted dual chamber microbial fuel cell with a CuInS$_2$ photocathode. *RSC Adv.* **2014**, *4*, 23790–23796. [CrossRef]
20. Kim, H.; Lee, K.; Razzaq, A.; Lee, S.H.; Grimes, C.A.; In, S. Photocoupled bioanode: A new approach for improved microbial fuel cell performance. *Energy Technol.* **2018**, *6*, 257–262. [CrossRef]
21. Cheng, S.; Logan, B.E. Sustainable and efficient biohydrogen production via electrohydrogenesis. *Proc. Natl. Acad. Sci. USA* **2007**, *104*, 18871–18873. [CrossRef] [PubMed]
22. Tice, R.C.; Kim, Y. Methanogenesis control by electrolytic oxygen production in microbial electrolysis cells. *Int. J. Hydrogen Energy* **2014**, *39*, 3079–3086. [CrossRef]
23. Logan, B.E. Scaling up microbial fuel cells and other bioelectrochemical systems. *Appl. Microbiol. Biotechnol.* **2010**, *85*, 1665–1671. [CrossRef] [PubMed]
24. Jiang, Y.; Liang, P.; Liu, P.; Wang, D.; Miao, B.; Huang, X. A novel microbial fuel cell sensor with biocathode sensing element. *Biosens. Bioelectron.* **2017**, *94*, 344–350. [CrossRef] [PubMed]
25. Qiu, R.; Zhang, B.G.; Li, J.X.; Lv, Q.; Wang, S.; Gu, Q. Enhanced vanadium (V) reduction and bioelectricity generation in microbial fuel cells with biocathode. *J. Power Sources* **2017**, *359*, 379–383. [CrossRef]
26. Rago, L.; Cristiani, P.; Villa, F.; Zecchin, S.; Colombo, A.; Cavalca, L.; Schievano, A. Influences of dissolved oxygen concentration on biocathodic microbial communities in microbial fuel cells. *Bioelectrochemistry* **2017**, *116*, 39–51. [CrossRef] [PubMed]
27. Jeremiasse, A.W.; Hamelers, H.V.; Buisman, C.J. Microbial electrolysis cell with a microbial biocathode. *Bioelectrochemistry* **2010**, *78*, 39–43. [CrossRef] [PubMed]
28. Jafary, T.; Daud, W.R.W.; Ghasemi, M.; Kim, B.H.; Md Jahim, J.; Ismail, M.; Lim, S.S. Biocathode in microbial electrolysis cell; present status and future prospects. *Renew. Sustain. Energy Rev.* **2015**, *47*, 23–33. [CrossRef]

29. Jafary, T.; Wan Daud, W.R.; Ghasemi, M.; Abu Bakar, M.H.; Sedighi, M.; Kim, B.H.; Carmona-Martínez, A.A.; Jahim, J.M.; Ismail, M. Clean hydrogen production in a full biological microbial electrolysis cell. *Int. J. Hydrogen Energy* **2018**. [CrossRef]
30. Huang, L.; Jiang, L.; Wang, Q.; Quan, X.; Yang, J.; Chen, L. Cobalt recovery with simultaneous methane and acetate production in biocathode microbial electrolysis cells. *Chem. Eng. J.* **2014**, *253*, 281–290. [CrossRef]
31. Chen, Y.; Shen, J.; Huang, L.; Pan, Y.; Quan, X. Enhanced Cd(II) removal with simultaneous hydrogen production in biocathode microbial electrolysis cells in the presence of acetate or NaHCO$_3$. *Int. J. Hydrogen Energy* **2016**, *41*, 13368–13379. [CrossRef]
32. Shen, R.; Liu, Z.; He, Y.; Zhang, Y.; Lu, J.; Zhu, Z.; Si, B.; Zhang, C.; Xing, X.-H. Microbial electrolysis cell to treat hydrothermal liquefied wastewater from cornstalk and recover hydrogen: Degradation of organic compounds and characterization of microbial community. *Int. J. Hydrogen Energy* **2016**, *41*, 4132–4142. [CrossRef]
33. Escapa, A.; Mateos, R.; Martínez, E.J.; Blanes, J. Microbial electrolysis cells: An emerging technology for wastewater treatment and energy recovery. From laboratory to pilot plant and beyond. *Renew. Sustain. Energy Rev.* **2016**, *55*, 942–956. [CrossRef]
34. Cusick, R.D.; Bryan, B.; Parker, D.S.; Merrill, M.D.; Mehanna, M.; Kiely, P.D.; Liu, G.; Logan, B.E.J.A.M. Biotechnology, Performance of a pilot-scale continuous flow microbial electrolysis cell fed winery wastewater. *Appl. Microbiol. Biotechnol.* **2011**, *89*, 2053–2063. [CrossRef] [PubMed]
35. Heidrich, E.S.; Dolfing, J.; Scott, K.; Edwards, S.R.; Jones, C.; Curtis, T.P.J.A.M. Biotechnology, Production of hydrogen from domestic wastewater in a pilot-scale microbial electrolysis cell. *Appl. Microbiol. Biotechnol.* **2013**, *97*, 6979–6989. [CrossRef] [PubMed]
36. Heidrich, E.S.; Edwards, S.R.; Dolfing, J.; Cotterill, S.E.; Curtis, T.P. Performance of a pilot scale microbial electrolysis cell fed on domestic wastewater at ambient temperatures for a 12 month period. *Bioresour. Technol.* **2014**, *173*, 87–95. [CrossRef] [PubMed]
37. Brown, R.K.; Harnisch, F.; Wirth, S.; Wahlandt, H.; Dockhorn, T.; Dichtl, N.; Schröder, U. Evaluating the effects of scaling up on the performance of bioelectrochemical systems using a technical scale microbial electrolysis cell. *Bioresour. Technol.* **2014**, *163*, 206–213. [CrossRef] [PubMed]
38. Cotterill, S.E.; Dolfing, J.; Jones, C.; Curtis, T.P.; Heidrich, E.S. Low Temperature Domestic Wastewater Treatment in a Microbial Electrolysis Cell with 1 m^2 Anodes: Towards System Scale-Up. *Fuel Cells* **2017**, *17*, 584–592. [CrossRef]
39. Cotterill, S.E.; Dolfing, J.; Curtis, T.P.; Heidrich, E.S. Community Assembly in Wastewater-Fed Pilot-Scale Microbial Electrolysis Cells. *Front. Energy Res.* **2018**, *6*. [CrossRef]
40. Mehanna, M.; Kiely, P.D.; Call, D.F.; Logan, B.E. Microbial Electrodialysis Cell for Simultaneous Water Desalination and Hydrogen Gas Production. *Environ. Sci. Technol.* **2010**, *44*, 9578–9583. [CrossRef] [PubMed]
41. Kim, Y.; Logan, B.E. Hydrogen production from inexhaustible supplies of fresh and salt water using microbial reverse-electrodialysis electrolysis cells. *Proc. Natl. Acad. Sci. USA* **2011**, *108*, 16176–16181. [CrossRef] [PubMed]
42. Li, X.; Jin, X.; Zhao, N.; Angelidaki, I.; Zhang, Y. Novel bio-electro-Fenton technology for azo dye wastewater treatment using microbial reverse-electrodialysis electrolysis cell. *Bioresour. Technol.* **2017**, *228*, 322–329. [CrossRef] [PubMed]
43. Cusick, R.D.; Logan, B.E. Phosphate recovery as struvite within a single chamber microbial electrolysis cell. *Bioresour. Technol.* **2012**, *107*, 110–115. [CrossRef] [PubMed]
44. Zamora, P.; Georgieva, T.; Ter Heijne, A.; Sleutels, T.H.J.A.; Jeremiasse, A.W.; Saakes, M.; Buisman, C.J.N.; Kuntke, P. Ammonia recovery from urine in a scaled-up Microbial Electrolysis Cell. *J. Power Sources* **2017**, *356*, 491–499. [CrossRef]
45. Almatouq, A.; Babatunde, A.O. Concurrent hydrogen production and phosphorus recovery in dual chamber microbial electrolysis cell. *Bioresour. Technol.* **2017**, *237*, 193–203. [CrossRef] [PubMed]
46. Chen, S.; Liu, G.; Zhang, R.; Qin, B.; Luo, Y. Development of the Microbial Electrolysis Desalination and Chemical-Production Cell for Desalination as Well as Acid and Alkali Productions. *Environ. Sci. Technol.* **2012**, *46*, 2467–2472. [CrossRef] [PubMed]
47. Ye, B.; Luo, H.; Lu, Y.; Liu, G.; Zhang, R.; Li, X. Improved performance of the microbial electrolysis desalination and chemical-production cell with enlarged anode and high applied voltages. *Bioresour. Technol.* **2017**, *244*, 913–919. [CrossRef] [PubMed]

48. Luo, H.; Li, H.; Lu, Y.; Liu, G.; Zhang, R. Treatment of reverse osmosis concentrate using microbial electrolysis desalination and chemical production cell. *Desalination* **2017**, *408*, 52–59. [CrossRef]
49. Ye, B.; Lu, Y.; Luo, H.; Liu, G.; Zhang, R. Tetramethyl ammonium hydroxide production using the microbial electrolysis desalination and chemical-production cell with long anode. *Bioresour. Technol.* **2018**, *251*, 403–406. [CrossRef] [PubMed]
50. Kim, Y.; Logan, B.E. Simultaneous removal of organic matter and salt ions from saline wastewater in bioelectrochemical systems. *Desalination* **2013**, *308*, 115–121. [CrossRef]
51. Carmona-Martínez, A.A.; Trably, E.; Milferstedt, K.; Lacroix, R.; Etcheverry, L.; Bernet, N. Long-term continuous production of H_2 in a microbial electrolysis cell (MEC) treating saline wastewater. *Water Res.* **2015**, *81*, 149–156. [CrossRef] [PubMed]
52. Shehab, N.A.; Ortiz-Medina, J.F.; Katuri, K.P.; Hari, A.R.; Amy, G.; Logan, B.E.; Saikaly, P.E. Enrichment of extremophilic exoelectrogens in microbial electrolysis cells using Red Sea brine pools as inocula. *Bioresour. Technol.* **2017**, *239*, 82–86. [CrossRef] [PubMed]
53. Hernández-Alonso, M.D.; Fresno, F.; Suárez, S.; Coronado, J.M.J.E.; Science, E. Development of alternative photocatalysts to TiO_2: Challenges and opportunities. *Energy Environ. Sci.* **2009**, *2*, 1231–1257. [CrossRef]
54. Henderson, M.A. A surface science perspective on TiO_2 photocatalysis. *Surf. Sci. Rep.* **2011**, *66*, 185–297. [CrossRef]
55. Sclafani, A.; Herrmann, J.M. Comparison of the Photoelectronic and Photocatalytic Activities of Various Anatase and Rutile Forms of Titania in Pure Liquid Organic Phases and in Aqueous Solutions. *J. Phys. Chem.* **1996**, *100*, 13655–13661. [CrossRef]
56. Angelome, P.C.; Andrini, L.; Calvo, M.E.; Requejo, F.G.; Bilmes, S.A.; Soler-Illia, G.J. Mesoporous anatase TiO_2 films: Use of Ti K XANES for the quantification of the nanocrystalline character and substrate effects in the photocatalysis behavior. *J. Phys. Chem. C* **2007**, *111*, 10886–10893. [CrossRef]
57. Colbeau-Justin, C.; Kunst, M.; Huguenin, D. Structural influence on charge-carrier lifetimes in TiO_2 powders studied by microwave absorption. *J. Mater. Sci.* **2003**, *38*, 2429–2437. [CrossRef]
58. Yamada, Y.; Kanemitsu, Y. Determination of electron and hole lifetimes of rutile and anatase TiO_2 single crystals. *Appl. Phys. Lett.* **2012**, *101*, 133907. [CrossRef]
59. Martini, I.; Hodak, J.H.; Hartland, G.V. Effect of Structure on Electron Transfer Reactions between Anthracene Dyes and TiO_2 Nanoparticles. *J. Phys. Chem. B* **1998**, *102*, 9508–9517. [CrossRef]
60. Roy, P.; Berger, S.; Schmuki, P. TiO_2 Nanotubes: Synthesis and Applications. *Angew. Chem. Int. Ed.* **2011**, *50*, 2904–2939. [CrossRef] [PubMed]
61. Lee, S.H.; Lee, K.-S.; Sorcar, S.; Razzaq, A.; Grimes, C.A.; In, S.-I. Wastewater treatment and electricity generation from a sunlight-powered single chamber microbial fuel cell. *J. Photochem. Photobiol. A Chem.* **2018**, *358*, 432–440. [CrossRef]
62. Melody, B.; Kinard, T.; Lessner, P.J.E. The Non-Thickness-Limited Growth of Anodic Oxide Films on Valve Metals. *Electrochem. Solid-State Lett.* **1998**, *1*, 126–129. [CrossRef]
63. Bard, A.; Parsons, R.; Jordan, J.J.N.Y. *Standard Potentials in Aqueous Solution*; CRC Press (Taylor & Francis): New York, NY, USA, 1985; ISBN 9780824772918.
64. Prakasam, H.E.; Shankar, K.; Paulose, M.; Varghese, O.K.; Grimes, C.A. A New Benchmark for TiO_2 Nanotube Array Growth by Anodization. *J. Phys. Chem. C* **2007**, *111*, 7235–7241. [CrossRef]
65. Regonini, D.; Bowen, C.R.; Jaroenworaluck, A.; Stevens, R. A review of growth mechanism, structure and crystallinity of anodized TiO_2 nanotubes. *Mater. Sci. Eng. R Rep.* **2013**, *74*, 377–406. [CrossRef]
66. Varghese, O.K.; Gong, D.; Paulose, M.; Grimes, C.A.; Dickey, E.C. Crystallization and high-temperature structural stability of titanium oxide nanotube arrays. *J. Mater. Res.* **2003**, *18*, 156–165. [CrossRef]
67. Liu, H.; Logan, B.E. Electricity generation using an air-cathode single chamber microbial fuel cell in the presence and absence of a proton exchange membrane. *Environ. Sci. Technol.* **2004**, *38*, 4040–4046. [CrossRef] [PubMed]
68. Cheng, S.; Liu, H.; Logan, B.E. Increased performance of single-chamber microbial fuel cells using an improved cathode structure. *Electrochem. Commun.* **2006**, *8*, 489–494. [CrossRef]
69. Cheng, S.; Logan, B.E. Ammonia treatment of carbon cloth anodes to enhance power generation of microbial fuel cells. *Electrochem. Commun.* **2007**, *9*, 492–496. [CrossRef]
70. Lovley, D.R.; Phillips, E.J.P. Novel mode of microbial energy metabolism: Organic carbon oxidation coupled to dissimilatory reduction of iron or manganese. *Appl. Environ. Microbiol.* **1988**, *54*, 1472–1480. [PubMed]

71. Gong, D.; Grimes, C.A.; Varghese, O.K.; Hu, W.; Singh, R.S.; Chen, Z.; Dickey, E.C. Titanium oxide nanotube arrays prepared by anodic oxidation. *J. Mater. Res.* **2001**, *16*, 3331–3334. [CrossRef]
72. Shankar, K.; Basham, J.I.; Allam, N.K.; Varghese, O.K.; Mor, G.K.; Feng, X.; Paulose, M.; Seabold, J.A.; Choi, K.-S.; Grimes, C.A. Recent advances in the use of TiO$_2$ nanotube and nanowire arrays for oxidative photoelectrochemistry. *J. Phys. Chem. C* **2009**, *113*, 6327–6359. [CrossRef]
73. Varghese, O.K.; Paulose, M.; Grimes, C.A. Long vertically aligned titania nanotubes on transparent conducting oxide for highly efficient solar cells. *Nat. Nanotechnol.* **2009**, *4*, 592–597. [CrossRef] [PubMed]
74. Sorcar, S.; Razzaq, A.; Tian, H.; Grimes, C.A.; In, S.-I. Facile electrochemical synthesis of anatase nano-architectured titanium dioxide films with reversible superhydrophilic behavior. *J. Ind. Eng. Chem.* **2017**, *46*, 203–211. [CrossRef]
75. Zhang, Y.; Zhu, W.; Cui, X.; Yao, W.; Duan, T. One-step hydrothermal synthesis of iron and nitrogen co-doped TiO$_2$ nanotubes with enhanced visible-light photocatalytic activity. *CrystEngComm* **2015**, *17*, 8368–8376. [CrossRef]
76. Ieropoulos, I.; Winfield, J.; Greenman, J. Effects of flow-rate, inoculum and time on the internal resistance of microbial fuel cells. *Bioresour. Technol.* **2010**, *101*, 3520–3525. [CrossRef] [PubMed]
77. Watson, V.J.; Logan, B.E. Analysis of polarization methods for elimination of power overshoot in microbial fuel cells. *Electrochem. Commun.* **2011**, *13*, 54–56. [CrossRef]
78. Kim, B.; An, J.; Chang, I.S. Elimination of power overshoot at bioanode through assistance current in microbial fuel cells. *ChemSusChem* **2017**, *10*, 612–617. [CrossRef] [PubMed]

© 2018 by the authors. Licensee MDPI, Basel, Switzerland. This article is an open access article distributed under the terms and conditions of the Creative Commons Attribution (CC BY) license (http://creativecommons.org/licenses/by/4.0/).

Article

The Role of Natural Laccase Redox Mediators in Simultaneous Dye Decolorization and Power Production in Microbial Fuel Cells

Priyadharshini Mani [1],*, Vallam Thodi Fidal Kumar [2], Taj Keshavarz [1], T. Sainathan Chandra [2] and Godfrey Kyazze [1],*

[1] Faculty of Science and Technology, University of Westminster, 115 New Cavendish Street, London W1W 6UW, UK; T.Keshavarz@westminster.ac.uk
[2] Department of Biotechnology, Indian Institute of Technology (Madras), Chennai 600036, India; vtfkbt@gmail.com (V.T.F.K.); chandrasainathan@gmail.com (T.S.C.)
* Correspondence: Priyadharshini.Mani@my.westminster.ac.uk (P.M.); G.Kyazze@westminster.ac.uk (G.K.)

Received: 1 November 2018; Accepted: 8 December 2018; Published: 10 December 2018

Abstract: Redox mediators could be used to improve the efficiency of microbial fuel cells (MFCs) by enhancing electron transfer rates and decreasing charge transfer resistance at electrodes. However, many artificial redox mediators are expensive and/or toxic. In this study, laccase enzyme was employed as a biocathode of MFCs in the presence of two natural redox mediators (syringaldehyde (Syr) and acetosyringone (As)), and for comparison, a commonly-used artificial mediator 2,2′-azinobis(3-ethylbenzthiazoline-6-sulfonic acid) (ABTS) was used to investigate their influence on azo dye decolorization and power production. The redox properties of the mediator-laccase systems were studied by cyclic voltammetry. The presence of ABTS and As increased power density from 54.7 ± 3.5 mW m^{-2} (control) to 77.2 ± 4.2 mW m^{-2} and 62.5 ± 3.7 mW m^{-2} respectively. The power decreased to 23.2 ± 2.1 mW m^{-2} for laccase with Syr. The cathodic decolorization of Acid orange 7 (AO7) by laccase indicated a 12–16% increase in decolorization efficiency with addition of mediators; and the Laccase-Acetosyringone system was the fastest, with 94% of original dye (100 mgL^{-1}) decolorized within 24 h. Electrochemical analysis to determine the redox properties of the mediators revealed that syringaldehyde did not produce any redox peaks, inferring that it was oxidized by laccase to other products, making it unavailable as a mediator, while acetosyringone and ABTS revealed two redox couples demonstrating the redox mediator properties of these compounds. Thus, acetosyringone served as an efficient natural redox mediator for laccase, aiding in increasing the rate of dye decolorization and power production in MFCs. Taken together, the results suggest that natural laccase redox mediators could have the potential to improve dye decolorization and power density in microbial fuel cells.

Keywords: acetosyringone; dye decolorization; laccase; natural redox mediators; power density; syringaldehyde

1. Introduction

Microbial fuel cells (MFCs) could have potential in treating dyeing effluents with simultaneous power production. At the cathode of MFCs, platinum and metal oxide catalysts are commonly used for the oxygen reduction reaction (ORR). In recent years oxidoreductase enzymes e.g., laccase, have been explored as cathode catalysts in MFCs as a possible alternative to platinum as a way of reducing the cost of materials needed to construct MFCs [1,2]. Laccase is a multi-copper containing enzyme that is capable of one electron oxidation of other substrates and four electron reduction of O_2 to H_2O [3,4]. The enzyme is widely utilised in the oxidation of phenolic and non-phenolic substrates such as dyes, pesticides, antibiotics etc.

The redox potential of the substrate should be lower than that of laccase for oxidation to be thermodynamically feasible. The redox potential range for fungal laccase is between 0.4–0.8 V vs. standard hydrogen electrode (SHE), which is suitable for oxidation of phenolic substrates; for non-phenolic substrates that have a redox potential of >1.3 V vs. SHE, and cannot be oxidized directly by laccase, a redox mediator is required [5]. A redox mediator is a small molecular weight compound that is oxidized by the enzyme and reduced by the substrate continuously. They act as electron shuttles for large substrates that cannot access the active site of the enzyme, e.g., due to steric hindrance [6]. In laccase mediator systems (LMS), the enzyme oxidizes the mediators to form stable radicals with high redox potential that diffuse away from the enzyme active site and oxidize the substrates and get reduced in the process. In this way, laccase indirectly oxidises substrates that have high redox potentials or large molecular sizes [7] (Figure 1).

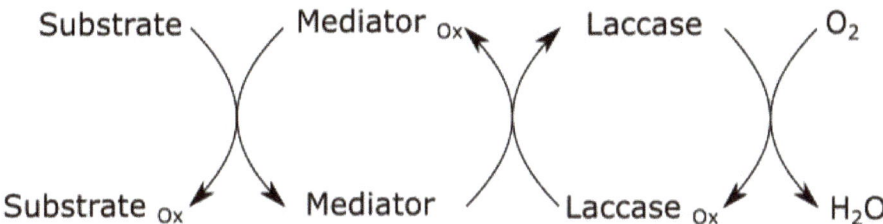

Figure 1. Laccase indirect substrate oxidation through mediators with oxygen as the final electron acceptor (Modified from [5]).

The first synthetic redox mediator reported was 2,2′-azinobis(3-ethylbenzthiazoline-6-sulfonic acid) (ABTS) for laccase from *Trametes versicolor* for oxidation of non-phenolic lignin compounds [8]. Another mediator that is involved in laccase lignin degradation and bleaching of kraft pulps is 1-Hydroxybenzotriazole (HBT) [9]. The LMS were initially used for delignification and bio-bleaching of wood pulps; nowadays, they are widely used for the degradation of xenobiotic compounds [10–12].

Although ABTS and HBT are the most widely used redox mediators for laccase, artificial mediators are not economically feasible, and they are toxic to the enzymes in the long run. In recent times, natural mediators have been explored for their environmental friendliness and low-cost. These natural mediators are phenolic compounds that exist in nature and mediate lignin oxidation in white rot fungi. Commonly-used phenolic mediators are syringaldehyde, acetosyringone, vanillin, methyl vanillate, p-coumaric acid, etc. [11]. The above mediators were compared with ABTS and HBT for the decolorization of different dyes with laccase from *T. versicolor* and *Pycnoporous cinnabarinus*. Syringaldehyde and acetosyringone showed 100% decolorization of Acid blue 74 with both laccases while there was >85% decolorization for Reactive Black 5 dye in less than 1 h for both the dyes. There was less than 50% decolorization of Acid Blue 74 by ABTS and HBT in the same time period. The phenolic mediators were more rapid and efficient in the oxidation of dyes than their synthetic counterparts [13]. The decolorization efficiency of various LMS depends on the dye structure. The mechanism of laccase mediation varies between each mediator. It was, for example, observed for lignin oxidation that ABTS/Laccase carried out alpha carbon oxidation and coupling of the lignin subunits, whereas HBT/laccase polymerized them [14].

Laccase stability and activity was decreased when incubated with artificial redox mediators ABTS, HBT, TEMPO (2,2,6,6-tetramethylpiperidin-1-yl)oxidanyl), and violuric acid (VA) at a concentration of 0.5 mM [14]. Even in the absence of any mediator, laccase activity decreased from 1000 U L^{-1} to 290 U L^{-1} in 15 days [15]. It is probably for this reason that high enzyme loadings such as 500 U mL^{-1} to 2000 U mL^{-1} are used in various dye decolorizing experiments [16,17]. Therefore methods to decrease the enzyme loading and reduce the cost of mediators should be further examined [18].

In this study natural mediators such as syringaldehyde and acetosyringone were studied with relatively low enzyme loadings (300 U L^{-1}) with a view to developing a low cost and sustainable laccase-mediator system. The use of laccase with natural phenolic mediators such as syringaldehyde and acetosyringone in a microbial fuel cell for dye decolorization has not been reported so far.

2. Materials and Methods

2.1. Chemicals

A crude commercial fungal laccase from *Trametes versicolor* with 10 U mg^{-1} of activity obtained from Enzyme India Pvt. Ltd., Chennai, India, was used. ABTS, syringaldehyde and acetosyringone were purchased from Sigma Aldrich, Cambridge, UK.

2.2. Operation of the Microbial Fuel Cell

An H-type MFC with a working volume of 200 mL in each chamber was used. Electrodes were made of non-woven carbon fiber, each with a surface area of 25 cm^2. Cation exchange membrane CMI7000 ion exchange membrane (Membranes International, Ringwood, NJ, USA) was soaked in 5% NaCl for 12 h before use. External resistance was 2000 Ω. This resistance was used because previous studies in our lab had found this to be the optimum for power production using the H-type reactor systems in our laboratory [19].

2.2.1. Anode Chamber Composition

The composition in the anode chamber was the same for all reactors. The anolyte consisted of minimal salts medium containing (per liter): 0.46 g NH_4Cl, 0.22 g $(NH_4)_2SO_4$, 0.117 g $MgSO_4$, 7.7 g $Na_2HPO_4.7H_2O$, 2.87 g NaH_2PO_4 along with 1% (v/v) trace minerals as described by Marsili et al. and 1% (v/v) vitamin mix as described by Wolin et al. [20,21]. The anolyte was supplemented with pyruvate as a carbon source at a concentration of 1 g L^{-1} and casein hydrolysate was also added at 500 mg L^{-1}. The pH of the anolyte was initially adjusted to 7. The anode was inoculated with 10% v/v *Shewanella oneidensis* MR-1 culture previously grown in Luria Bertani broth to an optical density of 0.4. The anode chamber was sparged for 10 min with nitrogen gas to remove any dissolved oxygen and to maintain an anaerobic environment.

2.2.2. Cathode Chamber Composition

The cathode chamber consisted of the commercial laccase from *Trametes versicolor* in 100 mM sodium acetate buffer solution (pH = 4.5) in the presence and absence of redox mediators. Laccase enzyme (300 U L^{-1}) was freely suspended in 200 mL of 100 mM acetate buffer (pH = 4.5) and 100 mg L^{-1} of Acid Orange 7 dye was added. After subsequent trial experiments, the concentration of the mediators were set at 50 µM. The cathode chamber was supplied with air through an air stone at a rate of 200 mL min^{-1}.

2.3. Experimental Design

Seven MFC systems were set up. System 1 was with *S. oneidensis* in the anode and laccase enzyme suspended in the cathode chamber in absence of mediators. This system is henceforth referred to as "Control Lac". System 2 was with *S. oneidensis* in the anode and laccase in the presence of ABTS in the cathode, hereafter referred to as "ABTS-lac". System 3 was with *S. oneidensis* in the anode and laccase in the presence of syringaldehyde in the cathode, hereafter referred to as "Syr-lac". System 4 was with *S. oneidensis* in the anode and laccase in the presence of acetosyringone in the cathode, hereafter referred to as "As-lac". System 5 was with *S. oneidensis* in the anode and syringaldehyde in the cathode without laccase, hereafter referred to as "Syringaldehyde". System 6 was with *S. oneidensis* in the anode and acetosyringone in cathode without laccase, hereafter to as "Acetosyringone". System 7 was with *S. oneidensis* in the anode and ABTS in the cathode without laccase, hereafter to as "ABTS". Experiments were conducted at a temperature of 30 °C.

2.4. Analytical Procedures

2.4.1. Acid Orange 7 Decolorization

The decolorization of AO7 at the cathode was measured at various time intervals using a UV-visible spectrophotometer at a wavelength of 484 nm, which is the maximum absorption wavelength for the dye. The decolorization efficiency (DE) was calculated by

$$DE\ (\%) = \frac{A_o - A_t}{A_o} \times 100$$

A_o and A_t are the absorbance units at the initial and each time point respectively. A time series was plotted for the absorbance values measured.

2.4.2. Electrochemical Analysis

The voltage across each MFC system was recorded at 10-min intervals using a Picolog data acquisition system (Pico Technology, St Neots, UK). The current through each system was calculated using Ohm's Law:

$$Current\ (I) = \frac{Voltage\ (V)}{Resistance\ (\Omega)}$$

The power produced was calculated using the following formula:

$$P = IV$$

where P is power in Watts, I is current in amperes and V is the electric potential in volts.

The power and current per surface area of anode electrode was used to calculate the power and current density. To carry out polarisation tests, each MFC unit was connected to various external resistances ranging from 10 Ω to 1 MΩ, and the steady state potential was measured using a multimeter.

2.4.3. Cyclic Voltammetry (CV) of Redox Mediators

The redox activity of the mediators ABTS, syringaldehyde, and acetosyringone in the presence and absence of laccase was analysed using cyclic voltammetry. A three-electrode system with the working electrode as glassy carbon, platinum as the counter, and Ag/AgCl as reference electrode was used. The three mediators were each added to 100 mM acetate buffer (pH 4.5) containing 0.3 U mL^{-1} laccase to give a final concentration of 50 µM. CV was carried out using a CH 660A potentiostat (CH Instruments) by cycling the potential between -1 V to 1 V at 50 mV s^{-1}.

2.4.4. Chronoamperometry (CA) of Laccase-Mediators

The effect of redox activity on current output was measured by CA. The same three-electrode system as used in Section 2.4.3 above was used. The working electrode was poised at 0.7 V, and laccase at a concentration of 0.3 U mL^{-1} was added to the 50 µM mediator solution to observe the change in current. CA was carried out using a CH 660A potentiostat (CH Instruments, Austin, TX, USA).

2.4.5. Statistical Analysis

All experimental data indicated in the text and graphs are the means of triplicate experiments unless otherwise stated. The error bars in the graphs and error values in the text represent standard deviations of the mean (SD). Data was analyzed using Microsoft Excel.

3. Results and Discussion

3.1. Power Generation

The power density was highest for the ABTS-lac system (77.2 ± 4.2 mW m^{-2}) compared to the control lac system (no mediators), which gave 54.7 ± 3.5 mW m^{-2} (Figure 2a). The power density of the ABTS-lac system is comparable to the performance of MFCs with platinum-coated cathodes (80 mW m^{-2}) obtained in a separate study in our laboratory [data submitted elsewhere for publication]. Similar trends were also obtained by Luo et al., when MFCs with laccase immobilized with Nafion-ABTS produced power (160 mW m^{-2}) equivalent to platinum-coated electrodes [1]. The power density in our study was much higher than Schaetzle et al., who obtained 37 mW m^{-2} with laccase-ABTS at the cathode of a MFC [22].

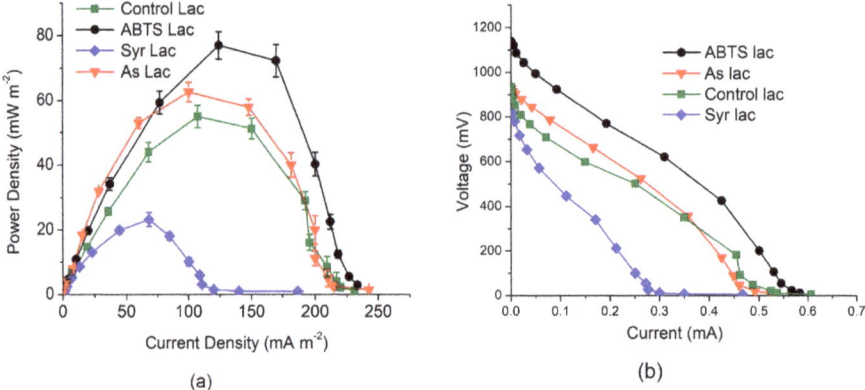

Figure 2. (a) Power density curves for mediator based laccase cathodes and control MFCs obtained by varying the external resistance from 10 Ω–1 MΩ; (b) Voltage vs. Current plot.

ABTS is oxidised by laccase, and it can be regenerated (reduced) by receiving electrons from the electrode and/or the dye [23]. The redox potential of the intermediates, ABTS$^+$ is 0.68 V and ABTS^{2+} is 1.09 V vs. SHE respectively [6]. The high redox potential of these ABTS radicals aids laccase in efficient reduction of oxygen which occurs at a potential of 1.2 V vs. SHE.

The As-lac system produced a P_{max} of 62.5 ± 3.7 mW m^{-2}, and for the Syr-lac system, it was 23.2 ± 2.1 mW m^{-2} (Figure 2a). The power density was higher for the As-lac than the control Lac system and vice versa for the Syr-lac system. The low power produced by the Syr-lac system is probably due to syringaldehyde acting as a substrate for laccase rather than a mediator. Electron donating groups of the benzene ring in phenolic compounds lowers their redox potentials which enables laccase to readily oxidize these substrates, the electrons released being used to reduce oxygen to water [12]. Since phenols are natural substrates for laccase, they are likely to be the source of electrons for oxidation rather than the cathode. This would reduce the power output of MFCs. The higher power density in the control lac systems indicates that in the absence of substrate oxidation, electrons are accepted from the cathode. Although acetosyringone is also a phenolic compound, the power density from the As-Lac system was greater than the control lac system, suggesting that the mediator could have been more efficiently regenerated compared to syringaldehyde. The detailed mechanism for the mediation is discussed in Section 3.3. Thus, from the power density data alone, it can be suggested that acetosyringone is a lower affinity substrate for laccase compared to syringaldehyde. This study is the first use of phenolic mediators in a MFC for laccase oxidation.

The internal resistance for the As-lac system was 1.5 kΩ compared to 1.8 kΩ for the control lac system; the ABTS-lac system had an internal resistance of 1.9 kΩ, while Syr-lac system had the highest resistance of 2.2 kΩ (Figure 2b).

In the absence of laccase, the power density for cathodes containing syringaldehyde and acetosyringone was 8.6 mW m^{-2} and 7.5 mW m^{-2} respectively.

3.2. Acid Orange 7 Decolorization

The decolorization rate of AO7 was highest in case of the As-lac system, followed by the Syr-lac system, and finally, the unmediated laccase biocathode (Figure 3). There was 94% decolorization in the As-lac system within 24 h of addition of dye. Decolorization in the control lac system was slightly slower with less than 80% decolorization in 24 h (Figure 3). Overall there was >95% decolorization for all laccase-based systems after 4 days which was not statistically significant compared to the control. Similar observations were observed for acetosyringone with Reactive Blue dye where >80% decolorization was observed in 2 h under non-MFC conditions [13]. As the two mediators are phenolic compounds that are substrates for laccase they are rapidly oxidized by the enzyme to produce phenoxy radicals that aid in dye decolorization [13]. In the presence of AO7 dye, the mediated laccase prefers the oxidation of dye for electrons rather than the anodic electron source with redox potentials of ca. −0.2 V vs. SHE [21]. The mediators are regenerated by abstraction of H$^+$ from the dye and e$^-$ from the cathode. Syringaldehyde and acetosyringone have been reported to have redox potentials of 0.660 V and 0.580 V vs. SHE respectively [24,25]. In the absence of laccase, the mediators have lower redox potential than AO7 (0.693 V vs. SHE); therefore, no decolorization was observed (Figure 3).

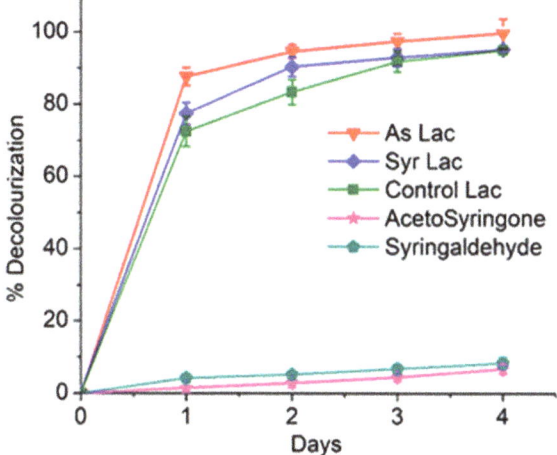

Figure 3. Decolorization of AO7 dye by laccase in the presence and absence of mediators over a period of 4 days.

Decolorization using the ABTS-Laccase system was also attempted for comparison but due to heavy interference with the color of ABTS in the presence of laccase (blue), the decolorization could not be studied effectively.

3.3. Electrochemical Activity of the Laccase Mediator Systems

To understand the reaction mechanism of the laccase-mediator systems, cyclic voltammetry was performed. The CV of the Syr system revealed a very weak oxidation peak at 0.73 V without any quantifiable cathodic current (Figure 4). In presence of laccase the oxidation peak was further decreased indicating syringadehyde's reduction reaction with the enzyme. There was absence of any redox peaks that are characteristic of redox mediators being regenerated. This might be due to laccase oxidizing syringaldehyde to syringic acid while producing phenoxy radicals and syringic acid further oxidizing to 2,6-dimethoxy-1,4-benzoquinone (DMBQ) [26,27] (Figure 5). Due to the subsequent oxidation of syringaldehyde, it is not regenerated and available as a mediator. Laccase is capable of catalyzing oxidative polymerization of quinone compounds to form polyhydroquinones [28].

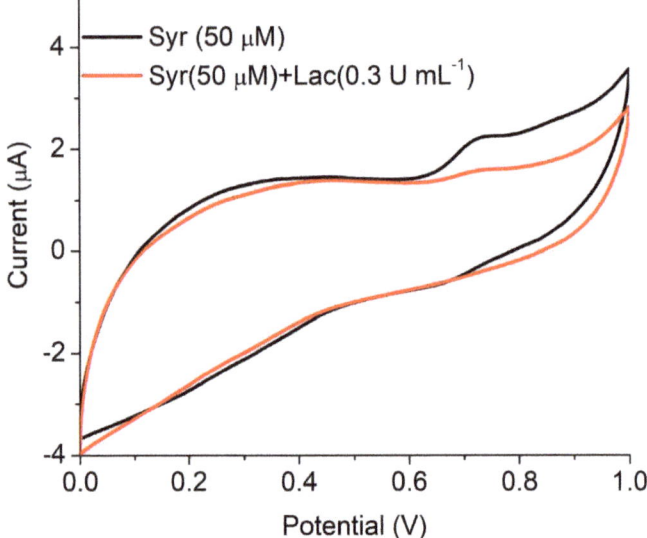

Figure 4. Cyclic voltammetry of syringaldehyde in the presence and absence of laccase at a scan rate of 50 mV s^{-1}. The potential indicated is vs. Ag/AgCl.

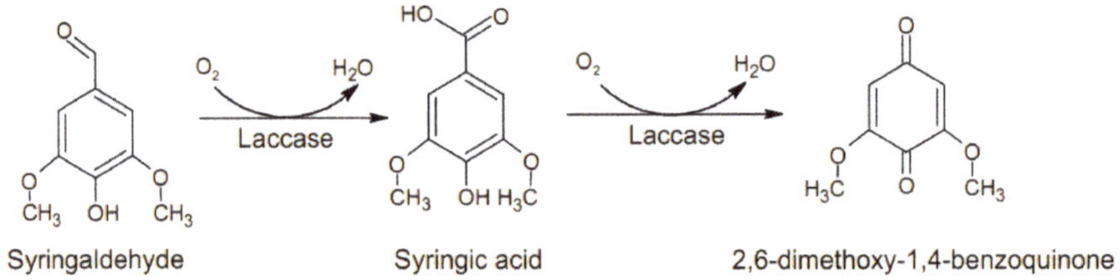

Figure 5. Laccase oxidation of syringaldehyde to syringic acid and subsequent oxidation to benzoquinones [29].

Therefore, the polymerization products of benzoquinone (DMBQ) formed could result in development of concentration gradients and mass transfer limitation at the electrode decreasing the power density. Another possible reason might be a result of the products inhibiting laccase enzyme activity. The CV of acetosyringone produced two redox couple peaks at 0.7 V/0.62 V and at 0.42 V/0.34 V (Figure 6). Acetosyringone has two major sites for oxidation/reduction reactions: a hydroxyl group at para position, and a keto group attached to the ring (Figure 7). The redox reactions at these two functional groups contribute to the redox couples in the CV. The functional groups are oxidized to form either a phenoxy radical or an enolate ion. These ions are intermediates of the oxidation reduction reaction stabilized by the aromatic ring. In presence of laccase, the peak at 0.7 V (close to laccase redox potential (0.780 V)) was reduced, whereas the cathodic current at the second redox peak was increased and shifted to 0.31 V. This indicates that one of the functional sites is preferably oxidized by laccase.

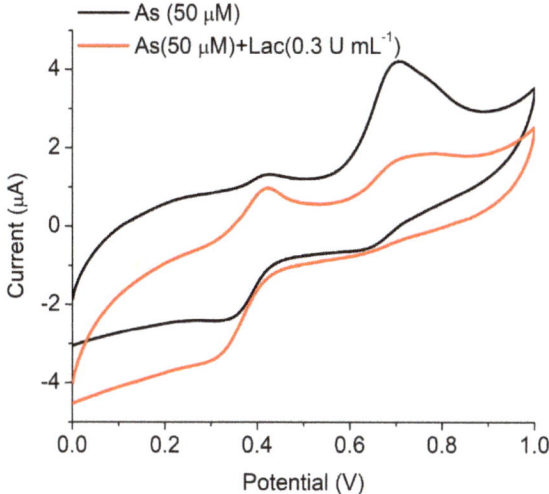

Figure 6. CV of acetosyringone indicating the oxidation/reduction peak in the presence and absence of laccase at a scan rate of 50 mV s^{-1}.

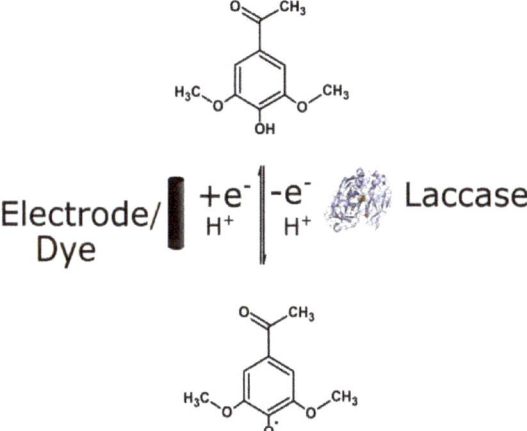

Figure 7. Electron transfer (ET) and Hydrogen atom transfer (HAT) oxidation mechanisms of acetosyringone mediated by laccase.

The two methods of mediation are the Electron transfer (ET) route and radical hydrogen atom transfer (HAT). In the ET route of mediation, only electrons are involved in the formation of free radicals and in the oxidation/reduction of the mediator. In the HAT mechanism, besides an electron, a H$^+$ ion is abstracted from hydroxyl groups of the mediators resulting in O free radical that aids in the mediation. From previous studies, it has been suggested that electron/hydrogen atom abstraction proceeds through the hydroxyl group present on the aromatic ring in acetosyringone [30]. Due to the presence of two functional groups, the mechanism of redox mediation in acetosyringone is a combination of HAT and ET route [26] (Figure 7). The presence of a keto group (as opposed to only hydroxyl) prevents laccase from completely oxidizing the substrate to a different product as observed in syringaldehyde. The acetosyringone is regenerated at the electrode/dye and is available as a mediator, contributing to the higher current output.

In case of ABTS, there were two redox couples: the first at 0.9 V/0.67 V, and the second at 0.63 V/0.45 V (Figure 8). The regeneration (reduction) of ABTS by accepting electrons from the electrode can be observed through increased cathodic current at 0.45 V in the presence of laccase. ABTS oxidation is a two-step mechanism where first it is oxidized to generate a cationic radical (ABTS$^+$) that is sequentially oxidized to a di-cation ABTS^{2+} (Figure 9) [31]. Hence, a clear decrease in both the oxidation peaks can be observed in the presence of laccase. ABTS is readily oxidized by laccase, and the mediator is constantly regenerated by accepting electrons from the electrode and the dye. The mechanism of ABTS mediation is through the electron transfer (ET) route between the enzyme and the substrate (Figure 9).

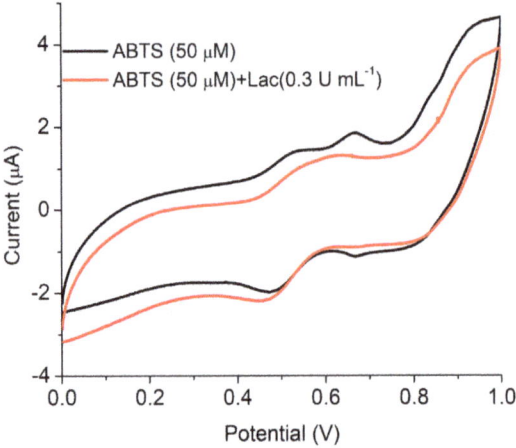

Figure 8. CV of ABTS indicating the oxidation/reduction peak in the presence and absence of laccase at a scan rate of 50 mV s^{-1}.

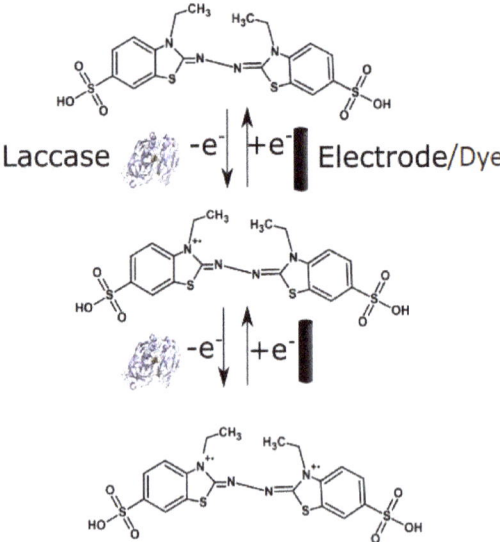

Figure 9. Two step oxidation/reduction of ABTS by laccase and electrode/dye respectively.

The effect of a laccase-mediator reaction on the current output was further tested by chronoamperometry. It was observed that ABTS system gave the highest cathodic current of 600 µA, whereas the syringaldehyde and acetosyringone systems produced 150 and 125 µA respectively (Figure 10).

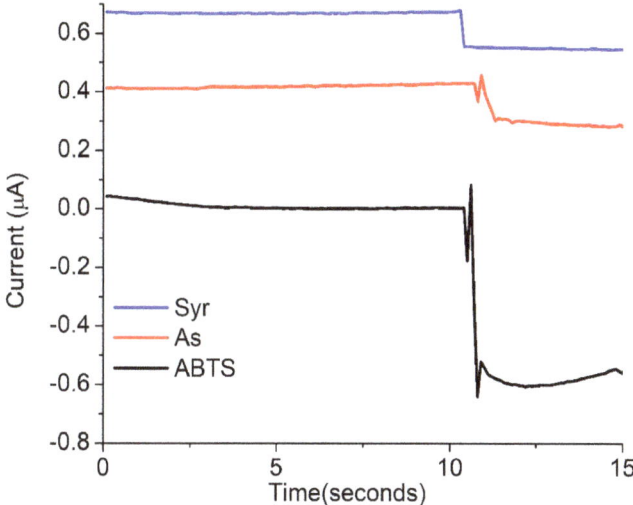

Figure 10. Chronoamperometry depicting the reduction current for each mediator at a concentration of 50 µM.

Although ABTS was the best mediator in terms of power production, acetosyringone performed comparably with an added advantage of dye decolorization. It is much cheaper and sustainable than ABTS. Syringaldehyde was the best substrate for laccase, and was completely oxidized; thus it did not act as a mediator to improve the performance of the MFC. Overall, the acetosyringone-lac system is preferred for dye decolorization and power production in a MFC.

4. Conclusions

Mediators could be used to improve the power density and efficiency of dye decolorization when used with laccase in the cathode of a MFC. For environmental and economic reasons, natural redox mediators such as syringaldehyde and acetosyringone are preferable, and in this study, their effectiveness was compared to those of a commonly used synthetic mediator ABTS. The presence of mediators increased the power density: the ABTS-lac system produced a P_{max} of 77.2 ± 4.2 mW m^{-2} while the As-lac system gave 62.5 ± 3.7 mW m^{-2}. The control lac system produced 54.7 ± 3.5 mW m^{-2} while the power density was the lowest for the Syr-lac system (23.2 ± 2.1 mW m^{-2}.) There was a 16% increase in decolorization efficiency with addition of mediators as compared to laccase in absence of mediators with As-lac achieving 94% decolorization in 24 h. The color of ABTS interfered with attempts to quantify the decolorization efficiency of the dye due to ABTS mediation and this is a limitation of this study. Electrochemical analysis performed to determine the redox properties of the mediators, revealed syringaldehyde did not produce any redox peaks, suggesting that it was oxidized by laccase to other products, making it unavailable as a mediator, while acetosyringone and ABTS revealed two redox couples demonstrating the redox behavior of these compounds. Thus, acetosyringone served as an efficient mediator for laccase, aiding in increased rate of dye decolorization and power production in a MFC.

Author Contributions: P.M. did most of the laboratory work and wrote the initial draft of the paper. G.K., T.K. and T.S.C. helped with the experimental design and review of progress from time to time. V.T.F.K. helped with electrochemical analyses. G.K. edited the paper prior to submission. All authors contributed to proof-reading the paper.

Acknowledgments: Priya wishes to thank her family for sponsoring her Ph.D. studies.

Conflicts of Interest: The authors declare no conflict of interest.

References

1. Luo, H.; Jin, S.; Fallgren, P.H.; Park, H.J.; Johnson, P.A. A Novel Laccase-Catalyzed Cathode for Microbial Fuel Cells. *Chem. Eng. J.* **2010**, *165*, 524–528. [CrossRef]
2. Savizi, I.S.P.; Kariminia, H.R.; Bakhshian, S. Simultaneous Decolorization and Bioelectricity Generation in a Dual Chamber Microbial Fuel Cell Using Electropolymerized-Enzymatic Cathode. *Environ. Sci. Technol.* **2012**, *46*, 6584–6593. [CrossRef] [PubMed]
3. Galhaup, C.; Haltrich, D. Enhanced Formation of Laccase Activity by the White-Rot Fungus *Trametes Pubescens* in the Presence of Copper. *Appl. Microbiol. Biotechnol.* **2001**, *56*, 225–232. [CrossRef] [PubMed]
4. Alneyadi, A.H.; Rauf, M.A.; Ashraf, S.S. Oxidoreductases for the remediation of organic pollutants in water—A critical review. *Crit. Rev. Biotechnol.* **2018**, *38*, 971–988. [CrossRef] [PubMed]
5. Morozova, O.V.; Shumakovich, G.P.; Shleev, S.V.; Yaropolov, Y.I. Laccase-Mediator systems and their Applications: A Review. *Appl. Biochem. Microbiol.* **2007**, *43*, 523–535. [CrossRef]
6. Christopher, L.P.; Yao, B.; Ji, Y. Lignin Biodegradation with Laccase-Mediator Systems. *Front. Energy Res.* **2014**, *2*, 12. [CrossRef]
7. Kunamneni, A.; Ballesteros, A.; Plou, F.J.; Alcalde, M. Fungal Laccase—A Versatile Enzyme for Biotechnological Applications. In *Communicating Current Research and Educational Topics and Trends in Applied Microbiology*; Mendez-Vilas, A., Ed.; FORMATEX: Badajoz, Spain, 2007; pp. 233–245. ISBN 978-84-611-9422-3.
8. Bourbonnais, R.; Paice, M.G. Oxidation of Non-Phenolic Substrates. *FEBS Lett.* **1990**, *267*, 99–102. [CrossRef]
9. Call, H.P.; Mücke, I. History, Overview and Application of Mediated Lignolytic Systems, Especially Lacasse-Mediator-Systems (Lignozyme®-Process). *J. Biotechnol.* **1997**, *53*, 163–202. [CrossRef]
10. Wu, Y.; Teng, Y.; Li, Z.; Liao, X.; Luo, Y. Potential Role of Polycyclic Aromatic Hydrocarbons (PAHs) Oxidation by Fungal Laccase in the Remediation of an Aged Contaminated Soil. *Soil Biol. Biochem.* **2008**, *40*, 789–796. [CrossRef]
11. Zeng, S.; Qin, X.; Xia, L. Degradation of the Herbicide Isoproturon by Laccase-Mediator Systems. *Biochem. Eng. J.* **2017**, *119*, 92–100. [CrossRef]
12. Cañas, A.I.; Camarero, S. Laccases and Their Natural Mediators: Biotechnological Tools for Sustainable Eco-Friendly Processes. *Biotechnol. Adv.* **2010**, *28*, 694–705. [CrossRef] [PubMed]
13. Camarero, S.; Ibarra, D.; Martinez, M.J.; Martinez, A.T. Lignin-Derived Compounds as Efficient Laccase Mediators for Decolorization of Different Types of Recalcitrant Dyes. *Appl. Environ. Microbiol.* **2005**, *71*, 1775–1784. [CrossRef] [PubMed]
14. Hilgers, R.; Vincken, J.-P.; Gruppen, H.; Kabel, M.A. Laccase/Mediator Systems: Their Reactivity toward Phenolic Lignin Structures. *ACS Sustain. Chem. Eng.* **2018**, *6*, 2037–2046. [CrossRef] [PubMed]
15. Kurniawati, S.; Nicell, J.A. Efficacy of Mediators for Enhancing the Laccase-Catalyzed Oxidation of Aqueous Phenol. *Enzyme Microb. Technol.* **2007**, *41*, 353–361. [CrossRef]
16. Mendoza, L.; Jonstrup, M.; Hatti-Kaul, R.; Mattiasson, B. Azo Dye Decolorization by a Laccase/Mediator System in a Membrane Reactor: Enzyme and Mediator Reusability. *Enzyme Microb. Technol.* **2011**, *49*, 478–484. [CrossRef] [PubMed]
17. Stoilova, I.; Krastanov, A.; Stanchev, V. Properties of Crude Laccase from *Trametes versicolor* Produced by Solid-Substrate Fermentation. *Adv. Biosci. Biotechnol.* **2010**, *1*, 208–215. [CrossRef]
18. Mani, P.; Keshavarz, T.; Chandra, T.S.; Kyazze, G. Decolourisation of Acid orange 7 in a microbial fuel cell with a laccase-based biocathode: Influence of mitigating pH changes in the cathode chamber. *Enzyme Microb. Technol.* **2017**, *96*, 170–176. [CrossRef]
19. Fernando, E. Treatment of Azo Dyes in Industrial Wastewater Using Microbial Fuel Cells. Ph.D. Thesis, University of Westminster, London, UK, 2014.

20. Wolin, E.A.; Wolin, M.J.; Wolfe, R.S. Formation of Methane by Bacterial Extracts. *J. Biol. Chem.* **1963**, *238*, 2882–2886.
21. Marsili, E.; Baron, D.B.; Shikhare, I.D.; Coursolle, D.; Gralnick, J.A.; Bond, D.R. Shewanella Secretes Flavins That Mediate Extracellular Electron Transfer. *Proc. Natl. Acad. Sci. USA* **2008**, *105*, 6–11. [CrossRef]
22. Schaetzle, O.; Barrière, F.; Schröder, U. An Improved Microbial Fuel Cell with Laccase as the Oxygen Reduction Catalyst. *Energy Environ. Sci.* **2009**, *2*, 96–99. [CrossRef]
23. Le Goff, A.; Holzinger, M.; Cosnier, S. Recent Progress in Oxygen-Reducing Laccase Biocathodes for Enzymatic Biofuel Cells. *Cell. Mol. Life Sci.* **2015**, *72*, 941–952. [CrossRef] [PubMed]
24. Pardo, I.; Chanagá, X.; Vicente, A.I.; Alcalde, M.; Camarero, S. New Colorimetric Screening Assays for the Directed Evolution of Fungal Laccases to Improve the Conversion of Plant Biomass. *BMC Biotechnol.* **2013**, *13*, 90. [CrossRef] [PubMed]
25. Baker, C.J.; Mock, N.M.; Whitaker, B.D.; Hammond, R.W.; Nemchinov, L.; Roberts, D.P.; Aver'yanov, A.A. Characterization of Apoplast Phenolics: Invitro Oxidation of Acetosyringone Results in a Rapid and Prolonged Increase in the Redox Potential. *Physiol. Mol. Plant Pathol.* **2014**, *86*, 57–63. [CrossRef]
26. Lin, H.; Su, J.; Liu, Y.; Yang, L. Catalytic Conversion of Lignocellulosic Biomass to Value-Added Organic Acids in Aqueous Media. In *Application of Hydrothermal Reactions to Biomass Conversion. Green Chemistry and Sustainable Technology*; Jin, F., Ed.; Springer: Berlin/Heidelberg, Germany, 2014; p. 109.
27. Volkova, N.; Ibrahim, V.; Hatti-Kaul, R. Laccase Catalysed Oxidation of Syringic Acid: Calorimetric Determination of Kinetic Parameters. *Enzyme Microb. Technol.* **2012**, *50*, 233–237. [CrossRef] [PubMed]
28. Sun, X.; Bai, R.; Zhang, Y.; Wang, Q.; Fan, X.; Yuan, J.; Cui, L.; Wang, P. Laccase-Catalyzed Oxidative Polymerization of Phenolic Compounds. *Appl. Biochem. Biotechnol.* **2013**, *171*, 1673–1680. [CrossRef] [PubMed]
29. Lahtinen, M.; Kruus, K.; Heinonen, P.; Sipilam, J. On the reactions of two fungal laccases differing in their redox potential with lignin model compounds: Products and their rates of formation. *J. Agric. Food Chem.* **2009**, *57*, 8357–8365. [CrossRef] [PubMed]
30. Martorana, A.; Sorace, L.; Boer, H.; Vazquez-Duhalt, R.; Basosi, R.; Baratto, M.C. A Spectroscopic Characterization of a Phenolic Natural Mediator in the Laccase Biocatalytic Reaction. *J. Mol. Catal. B Enzym.* **2013**, *97*, 203–208. [CrossRef]
31. Bourbonnais, R.; Leech, D.; Paice, M. Electrochemical Analysis of the Interactions of Laccase Mediators with Lignin Model Compounds. *Biochim. Biophys. Acta* **1998**, *1379*, 381–390. [CrossRef]

© 2018 by the authors. Licensee MDPI, Basel, Switzerland. This article is an open access article distributed under the terms and conditions of the Creative Commons Attribution (CC BY) license (http://creativecommons.org/licenses/by/4.0/).

MDPI
St. Alban-Anlage 66
4052 Basel
Switzerland
Tel. +41 61 683 77 34
Fax +41 61 302 89 18
www.mdpi.com

Energies Editorial Office
E-mail: energies@mdpi.com
www.mdpi.com/journal/energies